Hypersonic Slender Body Aerodynamics

Aerospace Series

Hypersonic Slender Body Aerodynamics
Ethirajan Rathakrishnan

Aircraft System Classifications: A Handbook of Characteristics and Design Guidelines
Allan Seabridge and Mohammad Radaei

UAS Integration into Civil Airspace: Policy, Regulations and Strategy
Douglas M. Marshall

Introduction to UAV Systems, Fifth Edition
Paul G. Fahlstrom, Thomas J. Gleason, Mohammad H. Sadraey

Introduction to Flight Testing
James W. Gregory, Tianshu Liu

Foundations of Space Dynamics
Ashish Tewari

Essentials of Supersonic Commercial Aircraft Conceptual Design
Egbert Torenbeek

Design of Unmanned Aerial Systems
Mohammad H. Sadraey

Future Propulsion Systems and Energy Sources in Sustainable Aviation
Saeed Farokhi

Flight Dynamics and Control of Aero and Space Vehicles
Rama K. Yedavalli

Design and Development of Aircraft Systems, 3rd Edition
Allan Seabridge, Ian Moir

Helicopter Flight Dynamics: Including a Treatment of Tiltrotor Aircraft, 3rd Edition
Gareth D. Padfield CEng, PhD, FRAeS

Space Flight Dynamics, 2nd Edition
Craig A. Kluever

Performance of the Jet Transport Airplane: Analysis Methods, Flight Operations, and Regulations
Trevor M. Young

Small Unmanned Fixed-wing Aircraft Design: A Practical Approach
Andrew J. Keane, András Sóbester, James P. Scanlan

Advanced UAV Aerodynamics, Flight Stability and Control: Novel Concepts, Theory and Applications
Pascual Marqués, Andrea Da Ronch

Differential Game Theory with Application to Missiles and Autonomous Systems Guidance
Farhan A. Faruqi

Introduction to Nonlinear Aeroelasticity
Grigorios Dimitriadis

Introduction to Aerospace Engineering with a Flight Test Perspective
Stephen Corda

Aircraft Control Allocation
Wayne Durham, Kenneth A. Bordignon, Roger Beck

Remotely Piloted Aircraft Systems: A Human Systems Integration Perspective
Nancy J. Cooke, Leah J. Rowe, Winston Bennett Jr., DeForest Q. Joralmon

Theory and Practice of Aircraft Performance
Ajoy Kumar Kundu, Mark A. Price, David Riordan

Adaptive Aeroservoelastic Control
Ashish Tewari

The Global Airline Industry, 2nd Edition
Peter Belobaba, Amedeo Odoni, Cynthia Barnhart, hristos Kassapoglou

Introduction to Aircraft Aeroelasticity and Loads, 2nd Edition
Jan R.Wright, Jonathan Edward Cooper

Theoretical and Computational Aerodynamics
Tapan K. Sengupta

Aircraft Aerodynamic Design: Geometry and Optimization
András Sóbester, Alexander I J Forrester

Stability and Control of Aircraft Systems: Introduction to Classical Feedback Control
Roy Langton

Aerospace Propulsion
T.W. Lee

Civil Avionics Systems, 2nd Edition
Ian Moir, Allan Seabridge, Malcolm Jukes

Aircraft Flight Dynamics and Control
Wayne Durham

Modelling and Managing Airport Performance
Konstantinos Zografos, Giovanni Andreatta, Amedeo Odoni

Advanced Aircraft Design: Conceptual Design, Analysis and Optimization of Subsonic Civil Airplanes
Egbert Torenbeek

Design and Analysis of Composite Structures: With Applications to Aerospace Structures, 2nd Edition
Christos Kassapoglou

Aircraft Systems Integration of Air-Launched Weapons
Keith A. Rigby

Understanding Aerodynamics: Arguing from the Real Physics
Doug McLean

Aircraft Design: A Systems Engineering Approach
Mohammad H. Sadraey

Theory of Lift: Introductory Computational Aerodynamics in MATLAB/Octave
G.D. McBain

Sense and Avoid in UAS: Research and Applications
Plamen Angelov

Morphing Aerospace Vehicles and Structures
John Valasek

Spacecraft Systems Engineering, 4th Edition
Peter Fortescue, Graham Swinerd, John Stark

Unmanned Aircraft Systems: UAVS Design, Development and Deployment
Reg Austin

Visit www.wiley.com to view more titles in the Aerospace Series

Hypersonic Slender Body Aerodynamics

Ethirajan Rathakrishnan

WILEY

This edition first published 2025

Registered Office(s)
John Wiley & Sons, Inc., 111 River Street, Hoboken, NJ 07030, USA
John Wiley & Sons Ltd, New Era House, 8 Oldlands Way, Bognor Regis, West Sussex, PO22 8NQ, UK

For details of our global editorial offices, customer services, and more information about Wiley products visit us at www.wiley.com.

Wiley also publishes its books in a variety of electronic formats and by print-on-demand. Some content that appears in standard print versions of this book may not be available in other formats.

Library of Congress Cataloging-in-Publication Data Applied for:
Hardback: 9781394285631

Cover image: © SONGPHOL THESAKIT/Getty Image
Cover design: Wiley

Set in 9.5/12.5pt STIXTwoText by Straive, Chennai, India
Printed and bound by CPI Group (UK) Ltd, Croydon, CR0 4YY
C9781394285631_060125

The manufacturer's authorized representative according to the EU General Product Safety Regulation is Wiley-VCH GmbH, Boschstr. 12, 69469 Weinheim, Germany, e-mail: Product_Safety@wiley.com.

This book is dedicated to my parents,
Mr. Thammanur Shunmugam Ethirajan
and
Mrs. Aandaal Ethirajan

Contents

Preface

This book is developed to serve as text for a course on *Slender Body Aerodynamics* at the introductory level course for undergraduate and graduate level.

The basic aim of this book is to make a complete text covering the basic and applied aspects of slender body aerodynamics for students, engineers, and applied physicists. The philosophy followed in this book is that the subject of aerodynamic theory is covered combining the theoretical analysis, physical features, and the application aspects.

Considerable number of solved examples are given in all the chapters to fix the concepts introduced, and large number of exercise problems along with answers are listed at the end of these chapters to test the understanding of the material studied.

The book is organized in a logical manner and the topics are discussed in a systematic way, beginning with the basic aspects and then proceeding to the involved aspects of theory and application.

The selected references given at the end are hoped to be a useful guide for further study of the voluminous subject.

This book is the outgrowth of lectures presented by the author over a number of years, both at undergraduate and graduate level. The student, or reader, is assumed to have a background in the basic courses of fluid mechanics, gas dynamics, thermodynamics, and heat transfer. Advanced undergraduate students should be able to handle the subject material comfortably. Sufficient details have been included, so that the text can be used for self study. Thus, the book can be useful for scientists and engineers working in the field of aerodynamics in industries and research laboratories.

My sincere thanks to my undergraduate and graduate students in India and abroad, who are directly and indirectly responsible for the development of this book.

I thank Dr. S. M. Aravindh Kumar, Assistant Professor, Department of Aerospace Engineering, SRM Institute of Science and Technology, for checking the manuscript and suggesting some useful changes that made the manuscript impressive.

For instructors only, a companion Solutions Manual that contains typed solutions to all the end-of-chapter problems and lecture slides for the complete book are available from the publisher.

Chennai
January 2025

Ethirajan Rathakrishnan

About the Author

Ethirajan Rathakrishnan is a professor of aerospace engineering at the Indian Institute of Technology Kanpur, India. He is well known internationally for his research in the area of high-speed jets. The limit for the passive control of jets, called the Rathakrishnan Limit, is his contribution to the field of jet research, and the concept of *breathing blunt nose (BBN)*, which simultaneously reduces the positive pressure at the nose and increases the low pressure at the base, is his contribution to drag reduction at hypersonic speeds. Positioning the twin-vortex Reynolds number at around 5000, by changing the geometry from cylinder, for which the maximum limit for the Reynolds number for positioning the twin-vortex was found to be around 160, by von Karman, to flat plate, is his addition to vortex flow theory. He has published a large number of research articles in many reputed international journals. He is a fellow of many professional societies including the Royal Aeronautical Society. Rathakrishnan serves as the editor-in-chief of the *International Review of Aerospace Engineering* (IREASE) and *International Review of Mechanical Engineering* (IREME) journals. He has authored the following books: *Gas Dynamics*, 7th ed. (PHI Learning, New Delhi, 2020); Fundamentals of Engineering Thermodynamics, 2nd ed. (PHI Learning, New Delhi, 2005); *Fluid Mechanics: An Introduction*, 4th ed. (PHI Learning, New Delhi, 2021); *Gas Tables*, 3rd ed. (Universities Press, Hyderabad, India, 2012); *Theory of Compressible Flows* (Maruzen Co., Ltd. Tokyo, Japan, 2008); *Gas Dynamics Work Book*, 2nd ed. (Praise Worthy Prize, Napoli, Italy, 2013); *Elements of Heat Transfer* (CRC Press, Taylor & Francis Group, Boca Raton, Florida, USA, 2012); *Theoretical Aerodynamics* (John Wiley, New Jersey, USA, 2013); *High Enthalpy Gas Dynamics* (John Wiley & Sons Inc., 2015); *Dynamique Des Gaz* (Praise Worthy Prize, Napoli, Italy, 2015); and *Instrumentation, Measurements and Experiments in Fluids*, 2nd ed. (CRC Press, Taylor & Francis Group, Boca Raton, Florida, USA, 2017), *Helicopter Aerodynamics*, (PHI Learning, New Delhi, 2019); *Applied Gas Dynamics* 2nd ed. (John Wiley & Sons Inc., 2019), *Introduction to Aerospace Engineering – Basic Principles of Flight* (John Wiley, New Jersey, USA, 2021), *Encyclopedia of Fluid Mechanics*

(CRC Press, Taylor & Francis Group, Boca Raton, Florida, USA, 2022), Fluid and Thermal Dynamics Answer Bank for Engineers: The Concise Guide with Formulas and Principles for Students and Professionals (Brown Walker Press, FL, USA, 2023) and *Mind Power The Sixth Sense* (Routledge, Taylor & Francis Group, Boca Raton, Florida, USA, 2023).

Nomenclature

A	aspect ratio
a	speed of sound
ac	aerodynamic center
b	span
Btu	British thermal unit
c	chord
c_i	inner chord
c_o	outer chord
cg	center of gravity
c.p	center of pressure
C_D	drag coefficient
C_{D_L}	coefficient of drag due to lift
C_{D_f}	skin friction drag coefficient
C_f	skin friction coefficient
C_L	lift coefficient
C_M	pitching moment coefficient
C_N	normal force coefficient
C_p	coefficient of pressure
C_T	thrust coefficient
c_p	specific heat at constant pressure
c_v	specific heat at constant volume
d	diameter
D	drag
D_f	drag due to friction
D_L	drag due to lift
dB	decibel
g_0	gravitational acceleration
h	altitude
h	height

h	specific enthalpy
H	total enthalpy
ISA	International Standard Atmosphere
K	hypersonic similarity parameter
Kn	Knudsen number
L	lift
L/D	aerodynamic efficiency
M	Mach number
M_{det}	shock detachment Mach number
M	moment
$\overline{M}$	mean molecular weight
m	molecular mass of air
n	number of degree of freedom
p	pressure
p_0	stagnation pressure
q	dynamic pressure ($\frac{1}{2}\rho V^2$)
q	source strength/length
$\overline{q_F}$	average rate at which heat is transferred to the nose per unit frontal area of the body
q_r	surface radiative heat flux
R	gas constant
R_c	radius of curvature
R_u	universal gas constant
$\mathrm{R_e}$	Reynolds number
s	span
S	wing area
S_{base}	base area (reference area)
S_{wet}	wetted area
St_o	Stanton number
T	temperature
T_0	stagnation temperature
t	time
V	velocity
$\mathbf{V}$	volume
$\overline{\mathbf{V}}$	nondimensionalized volume
u	perturbation velocity along the flow direction
u_g	normal component of velocity at any point on the body contour
v	perturbation velocity in the transverse direction
v_n	velocity component normal to the contour
w	perturbation velocity in the normal direction
x	flow direction

y	transverse direction
z	normal direction
z	compressibility factor

Greek symbols

α	angle of attack
α_{local}	local angle of attack
α	energy-accommodation coefficient
β	shock angle
Δ	"the change in"
ϵ	density ratio
ϵ	surface emissivity
Λ	leading-edge angle
λ	mean free path
λ_{eff}	slenderness ratio
Γ	circulation
γ	isentropic index
γ	specific heats ratio
γ	vorticity distribution
μ	viscosity coefficient
∇	total flux change
ν	Prandtl–Meyer function
Ω	angular velocity
Φ	velocity potential function
ϕ	perturbation velocity potential function
ϕ	shape correction factor
ρ	density
$\sigma(x)$	slope of shock
τ	volume parameter
$\overline{\tau_m}$	nondimensional thickness for the wing
θ	nose angle
θ	flow turning angle

Subscripts

0	stagnation state
∞	freestream
w	wall

About the Companion Website

The book is accompanied by a companion website.

www.wiley.com/go/rathakrishnan/hypersonic

The website contains:

- Lecture Slides
- Solution Manual

1

Basics

1.1 Introduction

Slender body configuration is a geometrical shape with one dimension much larger than the other two dimensions. Rocket shape is a typical example for slender body configuration. A slender wing shape is shown in Fig. 1.1.

For a slender wing, the height c is very large compared to the base s, that is, $c \gg s$, and the aspect ratio $\mathbf{A}$ (equal to the square of the wing span (s) divided by the wing area S) is much smaller than unity. That is,

$$\begin{aligned}\mathbf{A} &= \frac{\text{span}}{\text{chord}}\\ &= \frac{s}{c}\\ &= \frac{s \times s}{c \times s}\end{aligned}$$

or

$$\mathbf{A} = \frac{s^2}{S} \ll 1 \tag{1.1}$$

where $S = (c \times s)$ is the projected surface area of the wing normal to the flow direction. Equation (1.1) can be applied to geometries having aspect ratio as high as unity ($\mathbf{A} = 1$). Hence, $\mathbf{A} = 1$ may be taken as the upper limit. Such a configuration is used for supersonic civil transport aircraft.

Some of the typical slender body shapes, which are practically possible shapes and come under the study in this book are the Ogee and Gothic bodies as shown in Fig. 1.2.

The Ogee curve is a double S-shaped curvature along the cheeks that gives the face contour and dimension (a distinctive pattern with two continuous S-shaped curves narrowing at the nose and widening at the base, as in Fig. 1.2(a)). It is best

Hypersonic Slender Body Aerodynamics, First Edition. Ethirajan Rathakrishnan.

Companion Website: www.wiley.com/go/rathakrishnan/hypersonic

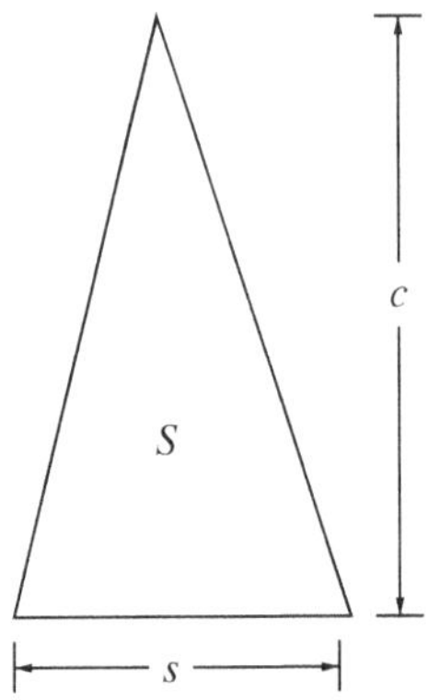

Figure 1.1 A slender wing.

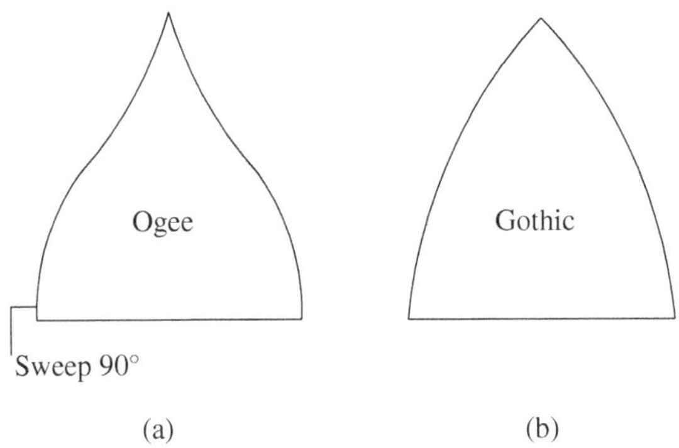

Figure 1.2 (a) Ogee shape and (b) Gothic shape.

appreciated when looking at a shape of the cheek[1] from a 45° angle. Gothic shape is a pointed arch, in which the two sides of the arch are joined by a sharp angle, as illustrated in Fig. 1.2(b).

Two typical rocket configurations, which are essentially slender bodies, are shown in Fig. 1.3. In the older type, the rockets had the control surfaces, namely, the fins only at the rear of it, as shown in Fig. 1.3(a). But in the new type of rockets, the control surface runs almost up to the nose, as shown in Fig. 1.3(b).

1 The Ogee curve is best seen when looking at a patient's cheek from an oblique angle. There is a convex fullness in the tail of the brow that gently curves where the upper and lower eyelids meet, then softly curves outward on the upper cheek to again curve inward just above the mouth. Younger women have a more pronounced Ogee curve, which enhances their appearance and looks more youthful. With age, the size of the curve diminishes leaving patients with flat cheeks and a lack of contour to the face. This is due to bone loss in the face as well as loss of volume, or fat, in the face. This theory originated with William Hogarth (18th-century English painter, satirist, and writer), and is an essential part of Hogarth's theory of aesthetics as described in his 1753 book *Analysis of Beauty*.

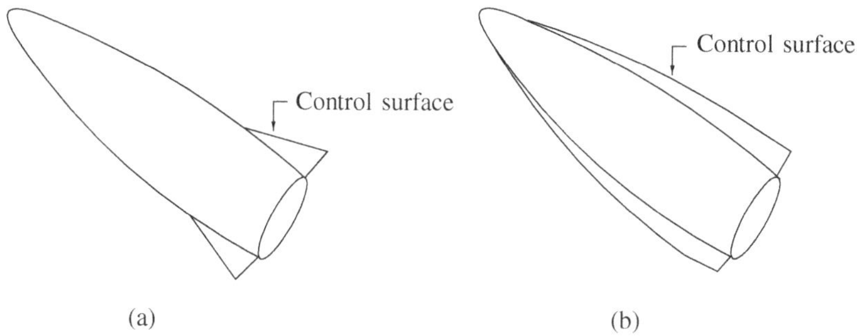

Figure 1.3 Rocket configuration: (a) Older type and (b) New type.

Example 1.1 If the aspect ratio of a triangular slender body of height 6 m is 0.3, determine the base length.

Solution

Given, $c = 6$ m, and $\mathbf{A} = 0.3$.
By Eq. (1.1),

$$\mathbf{A} = \frac{s^2}{S}$$

where the projected surface area $S = \frac{1}{2} \times s \times c$. Therefore, the base length, s, becomes

$$\mathbf{A} = \frac{s^2}{S}$$

$$\frac{s^2}{S} = 0.3$$

$$\frac{s^2}{\frac{1}{2} \times s \times c} = 0.3$$

$$s = \frac{0.3 \times c}{2}$$

$$= \frac{0.3 \times 6}{2}$$

$$= \boxed{0.9\,\text{m}}$$

1.2 Supersonic Transport Aircraft

The popular versions of supersonic transport aircraft are the Russian TU 144, the French-British Concorde, and Boeing 2707 of the United States of America.

Among these, TU 144 was designed to cruise at Mach 2.2, the Concorde at Mach 2, and the Boeing 2707 at Mach 2.7. Two prototype Concorde were made, one each in France and England. It's cruising altitude is 10 km, whereas the cruising altitude of Boeing 2707 is 13 km, which is the earth's curvature. Though TU 144 and Concorde aircraft were successfully tested and brought to practical use, the noise produced by them were around 120 dB, which is not acceptable to humans, especially when they fly at low altitudes, while approaching for landing or climbs after takeoff. This noise pollution added to the fuel economy issues forced the companies to ground them.

Currently, the American company Boom Technology, Inc. is developing supersonic transport aircraft.[2] Boom Technology is designing Mach 1.7, 65–88-passenger supersonic airliner. It is claimed that the maximum noise level of this aircraft is only around 60 dB. Therefore, this is expected to become a feasible supersonic passenger aircraft and come to commercial use in the year 2029.

Some agencies designed and tested hypersonic wings and wave riders, which are essentially slender body flying machines.

1.3 Wings of Supersonic Aircraft

The wings meant for supersonic flight should have the flow characteristics such that they are suitable for supersonic aircraft. To have an idea about this specific requirement, let us consider inviscid subsonic and supersonic flows over a flat plate of chord c at an angle of attack, as illustrated in Fig. 1.4.

If the flow is assumed to be inviscid, the plate in the subsonic flow would not experience any wave over it. Also, there won't be a skin friction acting over the body. Therefore, there won't be any drag acting on the plate. But, for the plate in the supersonic flow there will be compression and expansion waves at the leading

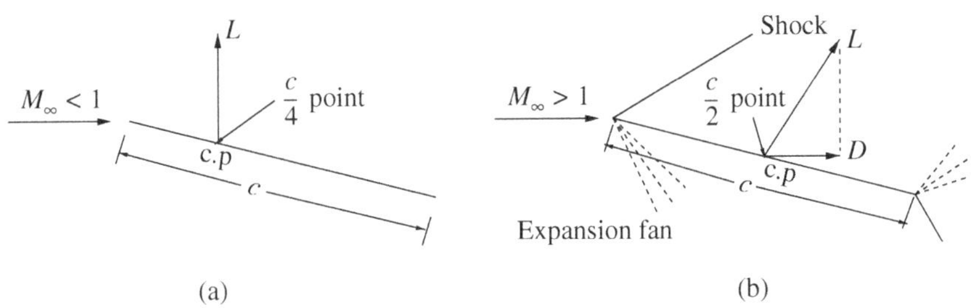

Figure 1.4 Flat plate in (a) subsonic flow and (b) supersonic flow.

---

2 Boom Technology, Inc. (trade name Boom Supersonic) is an American company designing a supersonic airliner named the Boom Overture.

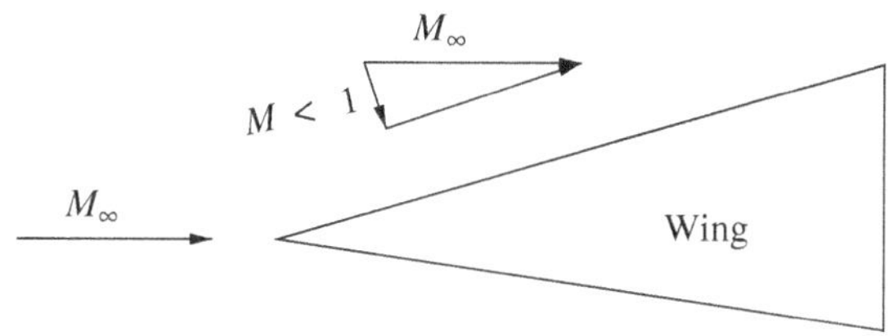

Figure 1.5 Flow past a slender wing.

and trailing edge, as illustrated in Fig. 1.4(b). Due to the presence of these waves, a drag force will be acting on the plate. This drag caused by the waves is termed as *wave drag*.[3] It is to be noted that the presence and absence of drag over the plate in supersonic (Fig. 1.4(b)) and subsonic (Fig. 1.4(a)) flows, respectively, are because the flow is assumed to be inviscid. But in actual situation, the flow will be viscous and, hence, the drag due to the skin friction[4] will be acting on the plate in subsonic and supersonic flows.

The center of pressure, c.p. (the point at which the line of action of the resultant of the forces due to pressure acts) is at the quarter chord point $c/4$ for the plate in subsonic stream and at the midpoint $c/2$ in the supersonic stream. Therefore, a slender body vehicle of large aspect ratio **A** flying at supersonic speed would encounter severe trim problems. But a slender wings' behavior is independent of freestream Mach number, therefore, there is no trim problem. It is the component of the freestream Mach number M_∞ that is normal to the leading edge, as shown in Fig. 1.5, that determines the behavior of the wing.

Designing slender wings having subsonic leading edges[5] is advantageous from an operation point of view. The advantage of this kind of leading edge is the following:

1. The center of pressure (the point of application of total pressure force exerted by the flow on the object), c.p., is independent of the freestream Mach number M_∞. That is, better trim.
2. No additional wave drag. In other words, because of the leading edge being subsonic, the wing has better drag characteristic.
3. Simple theory for design condition. This includes optimization of configuration.

---

3 Wave drag is a component of pressure drag due to compressibility effects. It is caused by the formation of shock waves around a body.
4 Skin friction drag is the drag caused by the fluid shearing along the surface of a body.
5 Wings with leading edge sweep greater than the Mach cone angle, so that the leading edges experience subsonic flow.

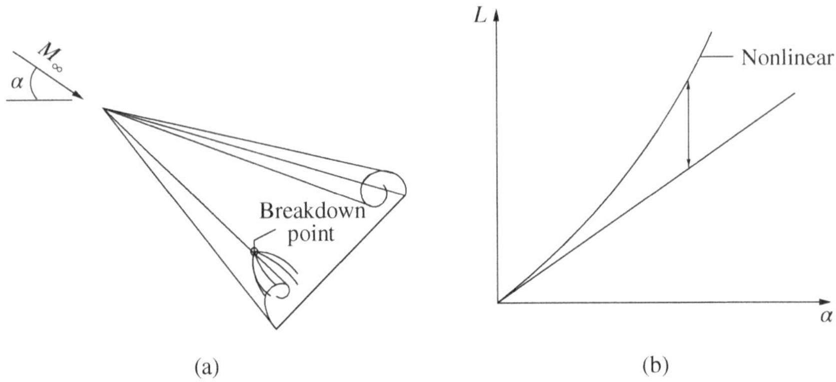

Figure 1.6 A slender wing at an angle of attack: (a) vortex sheets over the wing and (b) lift variation with angle of attack.

The disadvantage associated with the slender wings with subsonic leading edges is the following:

Off-design: In off-design conditions, there will be the leading edge separation, that is, when the angle of attack α is high, the flow has to go over the leading edge very quickly and it can no more go along the surface. The flow will get separated, curl and form small vortices behind the leading edge. These vortices will result in the formation of two *big vortex sheets* that would stay well above the upper surface, as illustrated in Fig. 1.6(a).

Due to the shedding of vortices and leading edge separation, the flow is attached and is different from a normal flow. The flow behavior is nonlinear, causing the lift variation with angle of attack as nonlinear, as illustrated in Fig. 1.6(b).

The vortex breakdown experienced is a stability problem. At some point, the vortex breaks down and this is a troublesome phenomenon because the breakdown point is far ahead of the trailing edge, as illustrated in Fig. 1.6(a). The area of the region over which the vortex breakdown takes place touches the upper surface of the wing. The flow behind this region is turbulent and the pressure distribution is changed fully. Therefore, it is necessary to make sure that the breakdown point is far behind the leading edge so that the area of this flow does not touch the wing's upper surface. Vortex breakdown point is very sensitive to yaw. There will be an additional unsymmetry introduced under yawed condition.

At the design condition, there will be no vortices when the flow hits the leading edge tangentially (this is achieved by cambering the wing).

Example 1.2 What should be the nose angle of a slender triangular wing flying at Mach 5 to have a subsonic leading edge.

Solution

Given, $M_\infty = 5$. For the triangular wing to have Mach number at its nose less than 1, the semi-nose angle $\theta/2$ the condition is $M_\infty \times \cos(\theta/2) \leq 1$, for hypersonic similarity. Therefore,

$$\begin{aligned} M_\infty \times \cos(\theta/2) &\leq 1 \\ 5 \times \cos(\theta/2) &\leq 1 \\ \cos(\theta/2) &\leq 1/5 \\ \theta/2 &\leq \cos^{-1}(0.2) \\ \theta/2 &\leq 78.46^\circ \end{aligned}$$

Thus, the nose angle has to be $\theta = 2 \times 78 = \boxed{156.92^\circ}$

Example 1.3 A slender delta wing of nose angle 10° cruises at supersonic speed. Determine the limiting speed up to which the wing will be shock wave-free.

Solution

The semi-angle of the nose (Fig. 1.5) is the angle by which the freestream flow with Mach number, M_∞, will be turned to make it flow parallel to the leading edge of the wing. For the wing to be free of shock waves, the flow Mach number, M, normal to its leading edge should be less than 1. That is, up to $M \leq 1$, the wing will be shock wave free. For the given wing, the semi-nose angle is 5°, therefore,

$$\begin{aligned} \frac{M}{M_\infty} &= \tan\,\theta \\ M_\infty &= \frac{M}{\tan\,\theta} \\ &= \frac{1}{\tan\,5} \\ &= \boxed{11.43} \end{aligned}$$

That is, for a freestream Mach number up to 11.43, the wing will fly without any shock wave over it.[6]

1.3.1 Sharp Leading Edge

Providing sharp leading edge to a slender wing with the aim of making the wing not to experience separation at the leading edge. This has to be done to ensure that the separation takes place always at the leading edge and is independent of

6 When the flow is supersonic, any change of flow property or direction is caused by waves. These waves are isentropic and nonisentropic compression waves (shock waves), expansion waves, and Mach waves. Among these, the compression and expansion waves can cause finite changes but the flow property changes caused by a Mach wave are insignificant [1].

the angle of attack. Separation line is fixed by this means, and the flow field is similar at all conditions.

Because of its application potential, the hypersonic flow past flat plate having sharp leading edge has been studied extensively. Unlike the low-speed counterpart, analytic solution valid for the entire extent of the surface is not accessible, only partitioning the flow into several sections, the flow can be analyzed. Experimental results show that, at the leading edge of the plate, a small noncontinuum region exits. This is followed by a continuum flow in which the boundary layer, generated by vorticity created at the surface, and "shock wave" or outer compressive zone are fully merged and indistinguishable from one another. This merged layer ultimately approached asymptotically the strong interaction region. It is here that the fully developed Rankine–Hugoniot shock wave interacts with the boundary layer so that, although these two viscous layers are distinct, their mutual developments are strongly coupled. A consistent matching between the unknown shock-wave location and boundary-layer displacement is required to describe the strong-interaction flow. A final regime, designated the weak interaction, occurs when the shock wave has degenerated so that ordinary boundary-layer theory and iterative techniques are applicable [2].

1.3.2 Ground Effect

The ground effect (the effect of aerodynamic buoyancy produced by a cushion of air below a vehicle moving close to the ground) encountered by a slender wing is strongly influenced by the aspect ratio. As the aspect ratio **A** increases, the ground effect increases.

Ascent speed and efficiency increase as the wings are brought closer due to a mirror-image "ground-effect" between the wings. Ground effect changes the drag and lift.

For a horizontal translation of the body near the curved ground, the attractive force will increase as the body is brought closer to the ground or as the absolute value of the angle of attack increases, and a convex (concave) ground will strengthen (weaken) the ground effect [3].

1.3.3 Viscous Effect

Viscous effects become very important because it makes the boundary layer to become so thick. This thick boundary layer causes large displacement thickness. This leads to the situation that the displacement effects cannot be neglected. Therefore, especially in hypersonic flow (also in supersonic flow), the interaction between the boundary layer and the outer flow should be considered. They cannot be treated separately. Viscous effects can have a substantial influence on the local bow shock structure surrounding high-speed slender bodies.

This coupling between the external inviscid flow and the viscous motion near the body surface is termed as hypersonic viscous interaction. Integrating the pressure induced by the flow passing over the body surface and the skin friction, due to viscosity, acting over the surface of the body, the viscous-interaction contribution to lift and drag acting were obtained [4].

1.4 Basic Equations

Let us consider the slender body in a flow of velocity $V(u, v, \text{and } w)$, as shown in Fig. 1.7. Assume that the flow is inviscid, steady, isentropic flow of a perfect gas. The unknown parameters to be solved are the velocity components u, v, and w along the x, y, and z directions, respectively, and the pressure p and the density ρ. The governing equations for this flow field are the Euler equations, continuity equation, and state equation. All these equations are potential. Therefore, for the given set of equations, the flow is irrotational. that is, curl $V = 0$.

For a potential flow, $V = \nabla\Phi$, where Φ is the velocity potential function. The five equations required for solving the unknowns: u, v, w, p, and ρ could be reduced to a single partial differential equation by using Φ. Using small perturbation assumption, this can be reduced to an ordinary differential equation in terms of the perturbation velocity potential function ϕ.

The potential function for the main stream is given by

$$\Phi_\infty = V_\infty x + \phi \tag{1.2}$$

where ϕ is the perturbation velocity potential and $\phi \ll V_\infty x$. With this, basic equation in terms of the linearized potential function will be

$$(1 - M_\infty^2)\,\phi_{xx} + \phi_{yy} + \phi_{zz} = 0 \tag{1.3}$$

This is the basic equation for all classical linearized theories.

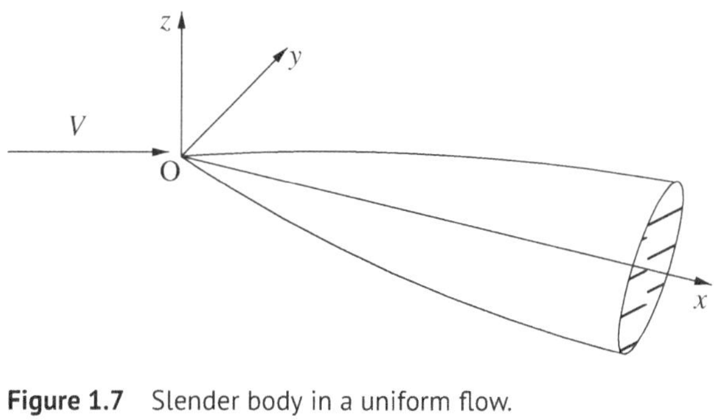

Figure 1.7 Slender body in a uniform flow.

Slender body: For slender bodies, the changes in velocity along the length of the body are much smaller than the velocities in the other directions. Therefore,

$$\phi_{xx} \ll \phi_{yy} + \phi_{zz}$$

For slender bodies, the basic equation becomes

$$\boxed{\phi_{yy} + \phi_{zz} = 0} \tag{1.4}$$

that is, the five equations required for solving the five unknowns: u, v, w, p, and ρ, are reduced to a single equation in terms of disturbance potential ϕ. This equation is simple to solve and the problem is a two-dimensional problem from the design point of view.

Example 1.4 A slender body cruises at Mach 8 at an altitude where the temperature is 200 K. If the perturbation velocity in the streamline direction is 10% of the freestream velocity, determine the perturbation velocity potential, treating the flow as one-dimensional.

Solution

Given, $M_\infty = 8$, and $T_\infty = 200\,\text{K}$.
The freestream speed of sound is

$$\begin{aligned} a_\infty &= \sqrt{\gamma R T\infty} \\ &= \sqrt{1.4 \times 287 \times 200} \\ &= 283.5\,\text{m/s} \end{aligned}$$

The freestream velocity is

$$\begin{aligned} V_\infty &= M_\infty a_\infty \\ &= 8 \times 283.5 \\ &= 2268\,\text{m/s} \end{aligned}$$

The freestream velocity potential is Φ_∞. Therefore,

$$\begin{aligned} \frac{\partial \Phi_\infty}{\partial x} &= V_\infty \\ &= 2268 \\ \Phi_\infty &= 2268x \end{aligned}$$

The perturbation velocity potential, ϕ, is $0.1\,\Phi_\infty$, therefore,

$$\begin{aligned} \phi &= (0.1 \times 2268)x \\ &= \boxed{226.8x} \end{aligned}$$

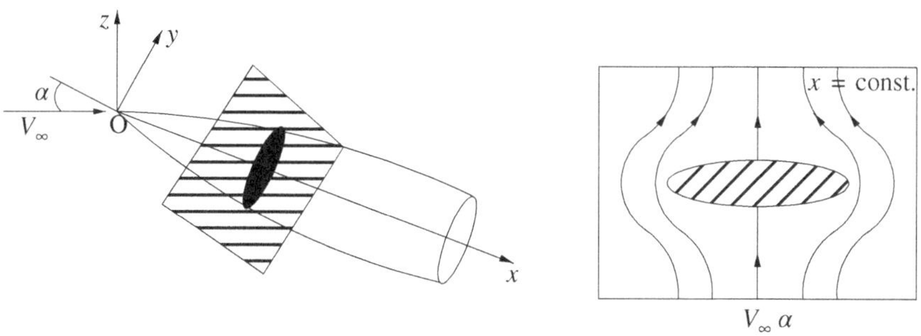

Figure 1.8 Flow past a slender body.

1.4.1 Velocity Potential Function in Slender Body Theory

Let us consider the flow past a slender body, as shown in Fig. 1.8.

Let ϕ be the velocity potential. The continuity equation, Eq. (1.4), is

$$\boxed{\phi_{yy} + \phi_{zz} = 0}$$

or

$$\Delta\phi = 0$$

The cross section depends on x. Therefore,

$$\phi = \phi(x, y, z)$$

The velocity field is

$$\vec{V} = i\,(V_\infty + u) + j\,v + k\,w$$

where V_∞ is the freestream velocity, and u, v, and w are the velocity components along x, y, and z directions, respectively. In terms of the velocity potential ϕ,

$$u = \frac{\partial \phi}{\partial x}$$

$$v = \frac{\partial \phi}{\partial y}$$

$$w = \frac{\partial \phi}{\partial z}$$

The boundary conditions are: $x = -\infty$, $u = 0$, $v = 0$, $w = V_\infty \alpha$.

The body contour is

$$z = z_b(x, y), \quad f(x, y, z) = z = z_b(x, y) - z_0$$

where subscript "b" refers to the body. On the body, the flow has to follow the surface tangentially, hence, there is no velocity normal to the surface, therefore,

$$\vec{n} \cdot \vec{V} = 0$$

Also, the perturbation velocity, u, in the streamline direction is very small compared to the freestream velocity, V_∞, that is $u \ll V_\infty$.

This gives rise to the kinematic flow condition: $\vec{n} \cdot \vec{V} = 0$ as

$$\boxed{V_\infty \frac{\partial z_b}{\partial x} + v \frac{\partial z_b}{\partial y} - w = 0} \tag{1.5}$$

on the body contour.

Example 1.5 A slender missile cruises at Mach 12 at 8000 m altitude. If the velocity perturbation in the axial, transverse, and normal direction are 5%, 1%, and 1% of the freestream velocity, respectively. Treating air as a perfect gas, find the perturbation potential and the incidence angle of the missile.

Solution

Given, $M_\infty = 12$, $h = 8000\,\text{m}$, $u = 0.05\,V_\infty$, $V = 0.01\,V_\infty$, and $w = 0.01\,V_\infty$. The specific heats ratio and gas constants for air are $\gamma = 1.4$, and $R = 287\,\text{m}^2/(\text{s}^2\,\text{K})$.

The temperature at 8000 m altitude, from atmospheric table, is $T_\infty = 236.15\,\text{K}$. This gives the freestream speed of sound as

$$\begin{aligned} a_\infty &= \sqrt{\gamma R T_\infty} \\ &= \sqrt{1.4 \times 287 \times 236.15} \\ &= 308\,\text{m/s} \end{aligned}$$

The freestream velocity is

$$\begin{aligned} V_\infty &= M_\infty a_\infty \\ &= 12 \times 308 \\ &= 3696\,\text{m/s} \end{aligned}$$

The perturbation velocity components are

$$\begin{aligned} u &= 0.05\,V_\infty \\ &= 0.05 \times 3696 \\ &= 184.8\,\text{m/s} \\ v &= 0.01\,V_\infty \\ &= 0.01 \times 3696 \\ &= 36.96\,\text{m/s} \\ w &= 0.01\,V_\infty \\ &= 0.01 \times 3696 \\ &= 36.96\,\text{m/s} \end{aligned}$$

Thus, the perturbation velocity potential is

$$\begin{aligned}\phi &= i(V_\infty + u) + j\,v + k\,w \\ &= (3696 + 184.8)\,i + 36.96j + 36.96\,k \\ &= \boxed{3880.8\,i + 36.96j + 36.96\,k}\end{aligned}$$

Also, $w = V_\infty\,\alpha$, where α is the angle of incidence. Therefore,

$$\begin{aligned}\alpha &= \frac{w}{V_\infty} \\ &= \frac{36.96}{3696} \\ &= 0.01 \text{ radian} \\ &= 0.01 \times 57.3 \\ &= \boxed{0.573^\circ}\end{aligned}$$

1.5 Pressure Distribution on the Surface

Examine the flow past from point 1 in the freestream flow, and point 2 over the slender body shown in Fig. 1.9. By energy equation,

$$h + \frac{1}{2}(\text{velocity})^2 = \text{constant} \tag{1.6}$$

where h is the specific enthalpy.

The pressure coefficient at any point on the body can be expressed as [5]

$$C_p = \frac{p - p_\infty}{\frac{1}{2}\rho_\infty V_\infty^2}$$

or

$$\boxed{C_p = 2\,\frac{u}{V_\infty} - \frac{v^2 + (w - V_\infty \alpha)^2}{V_\infty^2}} \tag{1.7}$$

where subscript ∞ refers to the freestream, u, v, and w are the perturbation velocity along the streamline, transverse, and normal direction, respectively, and α and V_∞ are the angle of attack and freestream velocity, respectively.

For a given geometry $z_b(x, y)$ and angle of attack α, the pressure coefficient $C_p(x, y)$ on the contour can be obtained as follows.

1. Find the potential function $\phi(x, y, z)$ from Eq. (1.4).
2. Find the velocity components from the potential function $\phi(x, y, z)$.
3. Calculate C_p from Eq. (1.7).

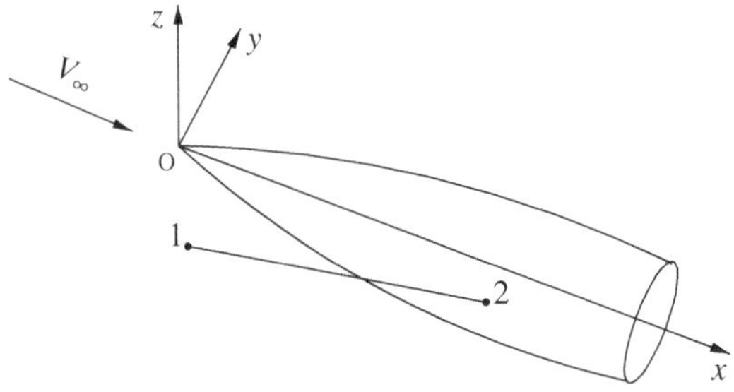

Figure 1.9 Flow past a slender body.

Remarks:

1. Equation (1.4) is a linear differential equation. Therefore, superposition of two solutions gives a new solution. However, this is not possible in the hypersonic flow. For such a case, one has to use nonlinear equations.
2. Equation (1.4) does not contain M_∞, that is, the flow is independent of the freestream Mach number.
3. Equation (1.4) does not contain x, therefore, an additional arbitrary function of x has to be used along with this equation.
4. This additional function of x should satisfy only the equation $u = \partial\phi/\partial x$. Even the boundary condition is independent of u. This additional function (of x only) cannot be derived from Eq. (1.4). This is a shortcoming in this procedure. To get this function, one has to go back to the original equation. Once this function is obtained, the original equation can be ignored and the present theory can be used. This function is called Raum einfluss (in German) or a *space influence*. Then, the velocity potential function becomes

$$\boxed{\phi = g(x, M_\infty)} \tag{1.8}$$

This will be the final solution.

1.5.1 General Solution

The potential function, $\phi(x, y, z)$, can be expressed as

$$\phi(x, y, z) = \phi_0(x, y, z) + \phi_\alpha(x, y, z) + g(x, M_\infty) \tag{1.9}$$

where $\phi_0(x, y, z)$ represents the longitudinal flow at $\alpha = 0°$ $(\Delta\phi_0 = 0)$, $\phi_\alpha(x, y, z)$ represents the transverse flow $(\alpha \neq 0°, \Delta\phi_\alpha = 0)$, and $g(x, M_\infty)$ represents the space influence.

Since $(\phi_0 + \phi_\alpha)$ satisfies the Laplace equation, the boundary conditions are also satisfied:

$$V_\infty \frac{\partial z_b}{\partial x} + v_0 \frac{\partial z_b}{\partial y} + v_\alpha \frac{\partial z_b}{\partial y} - w_0 - w_\alpha = 0 \tag{1.10}$$

This can again be split up into two equations, for 0 and α. Thus, the entire flow field can be considered as two. The longitudinal (thickness effect) and the transverse (angle of attack effect) flow problems can be considered separately. That is, ϕ_0 and ϕ_α can be uncoupled.

Example 1.6 A slender body that is similar to that shown in Fig. 1.9 flies at a freestream speed of Mach 3.0, with an angle of incidence of 2° at an altitude where the temperature is 220 K. If the perturbation velocity components along the x, y, and z directions are 10%, 3%, and 2% of the freestream velocity, respectively, at a particular location of the body, determine the pressure coefficient acting on the body at that point.

Solution

Given, $M_\infty = 3.0,\ \alpha = 2^\circ,\ T_\infty = 220\,\text{K},\ u = 0.1\,V_\infty,\ v = 0.03\,V_\infty,\ w = 0.02\,V_\infty$, and $\alpha = 2^\circ = 2/57.3 = 0.035$ radian.

The freestream speed of sound is

$$\begin{aligned} a_\infty &= \sqrt{\gamma R T_\infty} \\ &= \sqrt{1.4 \times 287 \times 220} \\ &= 297.3\,\text{m/s} \end{aligned}$$

The freestream speed is

$$\begin{aligned} V_\infty &= M_\infty a_\infty \\ &= 3 \times 297.3 \\ &= 891.9\,\text{m/s} \end{aligned}$$

The perturbation velocity components are

$$\begin{aligned} u &= 0.1\,V_\infty \\ &= 0.1 \times 891.9 \\ &= 89.19\,\text{m/s} \end{aligned}$$

$$\begin{aligned} v &= 0.03\,V_\infty \\ &= 0.03 \times 891.9 \\ &= 26.76\,\text{m/s} \end{aligned}$$

$$\begin{aligned} w &= 0.02\,V_\infty \\ &= 0.02 \times 891.9 \\ &= 17.84\,\text{m/s} \end{aligned}$$

The pressure coefficient, by Eq. (1.7), is

$$
\begin{aligned}
C_p &= 2\frac{u}{V_\infty} - \frac{v^2 + (w - V_\infty \alpha)^2}{V_\infty^2} \\
&= 2 \times \frac{89.19}{891.9} - \frac{26.76^2 + (17.84 - 891.9 \times 0.035)^2}{891.9^2} \\
&= 0.2 - \frac{26.76^2 + 13.38^2}{891.9^2} \\
&= 0.2 - \frac{716.1 + 179}{795485.61} \\
&= 0.2 - \frac{895.1}{795485.61} \\
&= 0.2 - 0.0011 \\
&= \boxed{0.199}
\end{aligned}
$$

1.5.2 Transverse Flow

The transverse flow field will be proportional to $V_\infty \alpha$, where V_∞ is the freestream velocity, and α is the angle of attack. The governing equation is

$$\boxed{\Delta\phi_\alpha = 0} \tag{1.11}$$

This represents lift problem. Also,

$$\frac{\partial z_b}{\partial y} = \frac{w_\alpha}{V_\alpha}$$

and

$$\boxed{v_n = 0} \tag{1.12}$$

This means that at every point on the body, the flow should be tangential to the body contour. Solution of this potential flow problem, together with the axial flow, will give the net flow tangential to the body contour.

1.5.3 Longitudinal Flow

The governing equation for a longitudinal flow at zero angle of attack, $\alpha = 0$, is

$$\boxed{\Delta\phi_0 = 0} \tag{1.13}$$

that is,

$$V_\infty \frac{\partial z_b}{\partial x} + v_0 \frac{\partial z_b}{\partial y} - w_0 = 0 \tag{1.14}$$

The velocity component tangential to the body contour is

$$\vec{V}_b \left(V_\infty + u_0, v_0, w_0\right)$$

The vector normal to the cross section $\vec{n}_{x\text{const.}} \left(0, \frac{\partial z_b}{\partial y}, -1\right)$ is

$$n_{x\text{const.}} = \sqrt{1 + \left(\frac{\partial z_b}{\partial y}\right)^2}$$

The component of velocity normal to the contour is

$$\begin{aligned} v_n &= -\frac{\vec{n}_{x\text{const.}}}{n_{x\text{const.}}} \cdot \vec{V}_b \\ &= \frac{-v_0 \frac{\partial z_b}{\partial y} + w_0}{\sqrt{1 + \left(\frac{\partial z_b}{\partial y}\right)^2}} \end{aligned}$$

that is,

$$\boxed{v_n = \frac{V_\infty \frac{\partial z_b}{\partial x}}{\sqrt{1 + \left(\frac{\partial z_b}{\partial y}\right)^2}}} \tag{1.15}$$

It is seen from Eq. (1.15) that the velocity normal to the surface depends only on the geometry of the body.

Now, the problem is essentially a two-dimensional problem. To this, the effect of $g(x, M_\infty)$ has to be added finally. This can be solved in the following two ways.

(i) Using complex functions which are holomorphic and then by conformal mapping.
(ii) With a source–sink distribution superposed vortex distribution. Here, the source–sink combination represents the longitudinal flow and the vortex stands for the transverse flow or lift problem. The lift problem gives the lift curve slope.

The potential function, $\phi(x, y, z)$, pressure distribution, C_p, on the contour and the total lift coefficient, C_L, for slender conical bodies of a few cross sections are listed in Figs. 1.10(a)–(f).

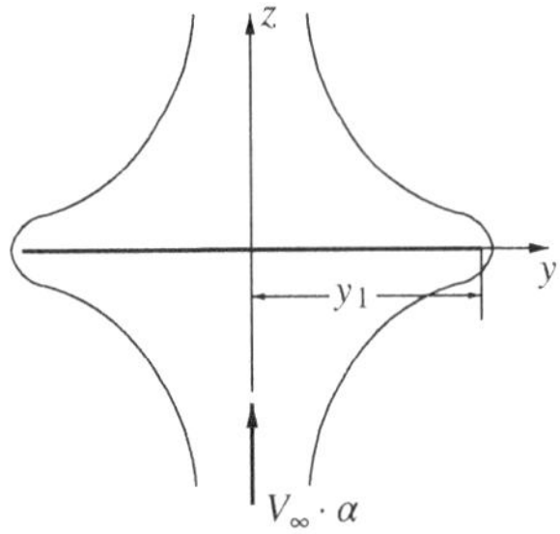

(a) Wing cross-section

Potential distribution:

$$\varphi(x, y, z) = \mathrm{Re}\left[V_\infty \cdot \alpha \cdot \sqrt{y_1^2 - z^{*2}}\right]$$

$$z^* = y + iz$$

Pressure distribution on the contour and total coefficient for conical bodies:

$$C_p = \mp \frac{\alpha \mathbf{A}}{2} \cdot \frac{1}{\sqrt{1 - (y/y_1)^2}}$$

$$C_L = \frac{\pi}{2} \mathbf{A}\, \alpha$$

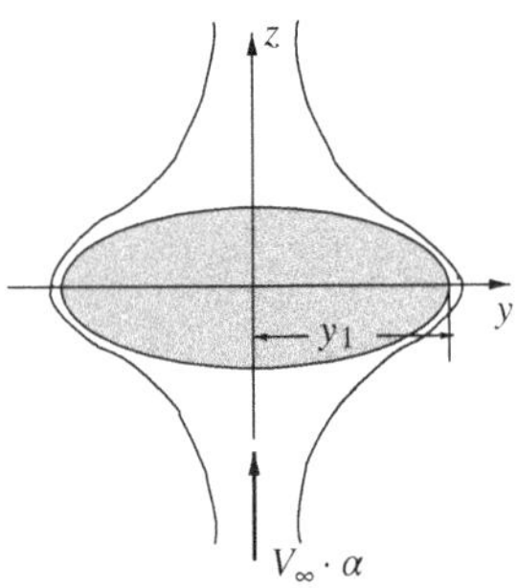

(b) Wing cross-section

Potential distribution:

$$\varphi(x, y, z) = \mathrm{Re}\left[- \frac{iV_\infty \cdot \alpha}{2} y_1 \left(\frac{z^*}{y_1} + \sqrt{\left(\frac{z^*}{y_1}\right)^2 + \delta^2 - 1} - \frac{(1+\delta)^2}{\frac{z^*}{y_1} + \sqrt{\left(\frac{z^*}{y_1}\right)^2 + \delta^2 - 1}} \right)\right]$$

$$z^* = y + iz, \quad \delta = z_1/y_1$$

Pressure distribution on the contour and total lift coefficient for conical bodies

$$C_p = \mp \frac{(1+\delta)\,\alpha\, \mathrm{A}}{2} \cdot \frac{\sqrt{1 - (y/y_1)^2}}{1 - (1 - \delta^2)(y/y_1)^2}$$

$$C_L = \frac{\pi}{2} \mathbf{A}\, \alpha$$

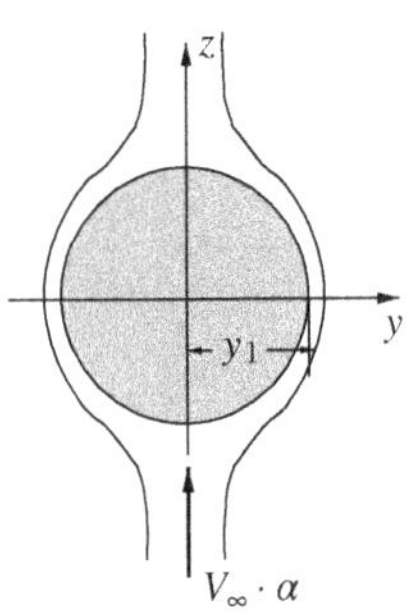

(c) Wing cross-section

Potential distribution:

$$\varphi(x, y, z) = \mathrm{Re}\left[- iV_\infty \cdot \alpha z^* \left(1 - \frac{R^2}{z^{*2}}\right)\right]$$

$$z^* = y + ix$$

Pressure distribution on the contour and total lift coefficient for conical bodies:

$$C_p = \mp \frac{(1+\delta)\,\alpha \mathbf{A}}{2} \sqrt{1 - \left(\frac{y}{y_1}\right)^2}$$

$$C_L = \frac{\pi}{2} \mathbf{A}\, \alpha$$

Figure 1.10 (a)–(f) Potential distribution and C_p and C_L for conical slender bodies.

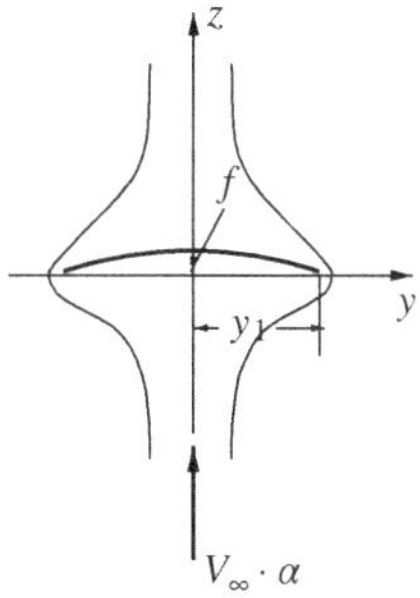

(d) Wing cross-section

Ref. J.H.B. Smith, The properties of thin conically cambered wing, according to slender body theory, ARC.R.M. No. 3135 (1960)

$$C_L = \frac{\pi}{2}\mathbf{A}\left(\alpha - \frac{5}{16}\mathbf{A}\frac{f}{y_1}\right),\quad \text{for } \left(\frac{f}{y_1}\right)^2 \ll 1 \text{ (small camber)}$$

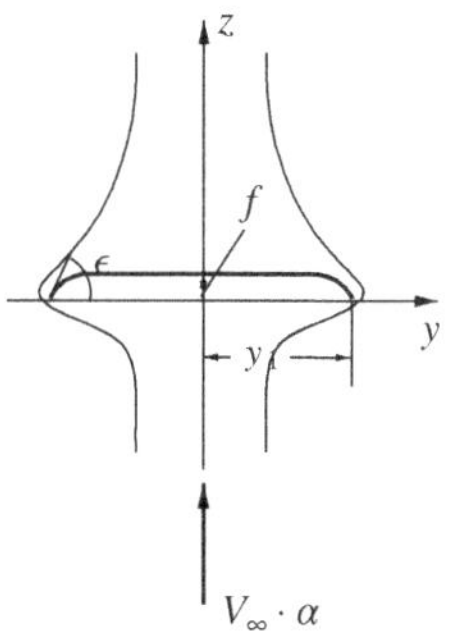

(e) Wing cross-section

See J.C. Cook, Properties of a two-parameter family of thin conically cambered delta wings by slender-body theroy, ARC.R.M. No. 3249 (1962)

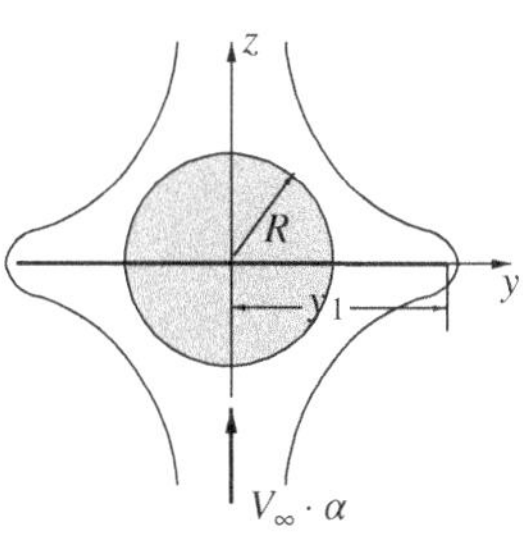

(f) Wing cross-section

Potential distribution:

$$\varphi(x, y, z) = \text{Re}\left[-iV_\infty \cdot \alpha \cdot \sqrt{\left(z^* + \left(\frac{R^2}{z}\right)^2 - \left(y_1 + \frac{R^2}{y_1}\right)^2\right.}\right]$$

$z^* = y + iz$

Body contour: $z^* \overline{z^*} = R^2$

Wing: $(R \le y \le y_1)$: $z^* = y$

Pressure distribution on the contour and total lift lift coefficient for conical bodes:

C_p: See J.N. Nielsen, Missile Aerodynamics, McGraw-Hill Book Co., Inc, NY

$$C_L = \frac{\pi}{2}\mathbf{A}\,\alpha\left[1 - \left(\frac{R}{y_1}\right)^2 + \left(\frac{R}{y_1}\right)^4\right]$$

Figure 1.10 (*Continued*)

1.6 Theoretical Methods – Inviscid Flow

Most of the early developments in supersonic missile aerodynamics started with the assumption of inviscid flow that was given a small perturbation by the movement of the missile. If the body were sufficiently long and thin, the "slender body" theory could be used to describe the perturbed flow. A linearized partial differential equation, the potential equation, described the flow field. The solution for this simplified model yields the estimates of dimensionless coefficients of lift (C_L) and drag due to lift (C_{D_L}) that are independent of the flight Mach number and body shape (within the slenderness restrictions of the theory) and of the static pitching moment (C_M) that depends on missile volume:

$$C_L = 2\alpha \tag{1.16}$$

$$C_{D_L} = C_L \frac{\alpha}{2} = \alpha^2 \tag{1.17}$$

$$C_M = -2\alpha \left(1 - \frac{\text{Volume}}{l\, S_{\text{base}}}\right) \tag{1.18}$$

where α is the angle of attack, l is the body length, S_{base} is the base area (reference area), and the reference center for C_M is the vertex of the nose. The wave drag can be estimated using the second derivatives of the axial distribution of the cross-sectional area.

Example 1.7 If a cylindrical missile of diameter 1 m, length 4 m, with a conical nose of height 3 m flies at 3° angle of attack, determine the lift, drag, and pitching moment acting on the missile, treating the flow as inviscid.

Solution

Given, $l = 4$ m, $d = 1$ m, cone height $h = 3$ m, and $\alpha = 3°$.

The volume of the missile is

$$\begin{aligned}
\mathbf{V} &= \mathbf{V}_{\text{cone}} + \mathbf{V}_{\text{cylinder}} \\
&= \left(\frac{1}{3} \times \pi \times 0.5^2 \times 3\right) + \left(\frac{\pi \times 1^2}{4} \times 4\right) \\
&= 0.785 + 3.141 \\
&= 3.926\,\text{m}^3
\end{aligned}$$

The lift, drag, and pitching moment coefficients, by Eqs. (1.16), (1.17), and (1.18), respectively, are

$$\begin{aligned}
C_L &= 2\alpha \\
&= 2 \times \frac{3}{57.3} \\
&= \boxed{0.1047}
\end{aligned}$$

$$C_D = \alpha^2$$
$$= \left(\frac{3}{57.3}\right)^2$$
$$= \boxed{0.00274}$$

$$C_M = -2\alpha\left(1 - \frac{\mathbf{V}}{l\,S_{\text{base}}}\right)$$
$$= -2 \times \frac{3}{57.3} \times \left(1 - \frac{3.926}{4 \times (\pi \times 0.5^2)}\right)$$
$$= -2 \times \frac{3}{57.3} \times \left(1 - \frac{3.926}{3.141}\right)$$
$$= -0.1047 \times (1 - 1.25)$$
$$= \boxed{0.026}$$

Example 1.8 For the slender wing without camber and thickness, whose side view, as shown in Fig. 1.11, find the expression for pressure coefficient.

On the contour: $z = 0$. The potential function for the flow is

$$\phi(x, y, 0) = V_\infty \cdot \alpha\sqrt{y_1^2 - y^2} \tag{1.19}$$

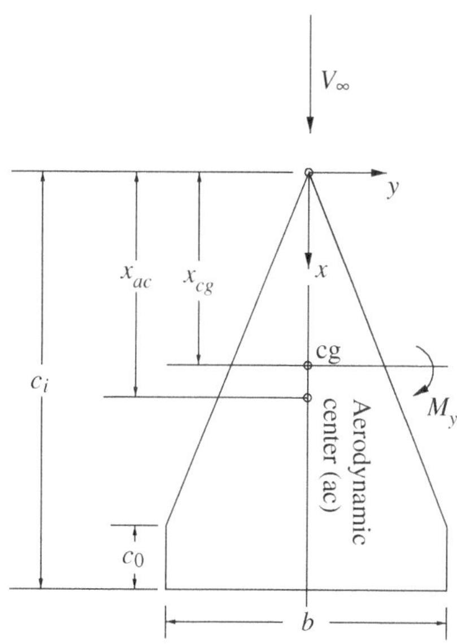

Figure 1.11 Transverse flow past a slender wing.

This is an elliptical potential distribution. The velocity is given by

$$u(x, y, 0) = \pm V_\infty \cdot \alpha \frac{y_1}{\sqrt{y_1^2 - y^2}} \frac{dy_1}{dx} \tag{1.20}$$

The positive sign (+ term) in Eq. (1.20) is for the upper surface and negative sign (– term) is for the lower surface.

The pressure coefficient C_p is given by

$$\boxed{C_p(x, y) = -\frac{2u}{V_\infty} \mp 2\alpha \frac{y_1}{\sqrt{y_1^2 - y^2}} \frac{dy_1}{dx}} \tag{1.21}$$

where the negative sign (– term) is for the upper surface and the positive sign (+ term) is for the lower surface. Thus, dy_1/dx determines the pressure distribution, that is, if the trailing edge gradient is zero the pressure there will be zero. For such a planform, the flow separates after the point where $dy_1/dx = 0$, as shown in Fig. 1.12. Therefore, the pressure corresponds to $dy_1/dx = 0$.

Examine the pressure coefficient variation over the slender body, as shown in Fig. 1.13. Near the leading edge the dimensionless pressure, that is the pressure coefficient goes to ∞ because of the sharp edge.

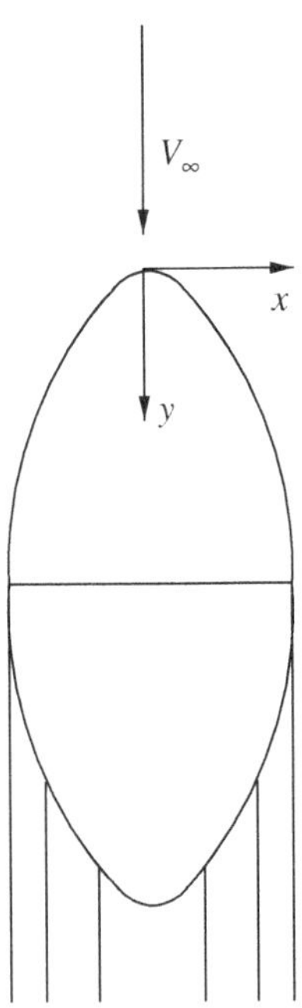

Figure 1.12 Flow separation from a slender body.

Figure 1.13 Pressure variation over the slender body.

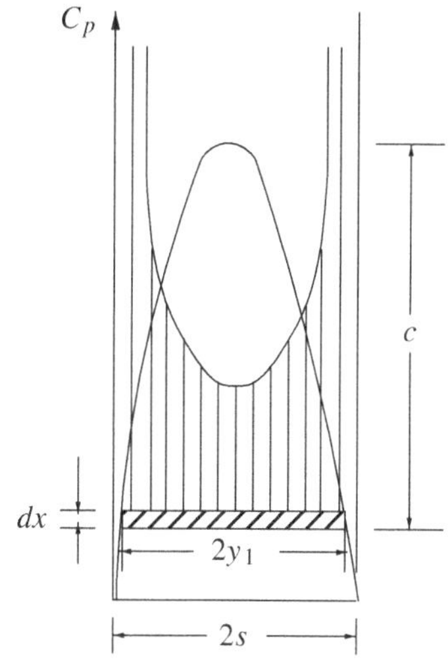

The lift acting over the section of thickness dx is given by

$$dL = \int_{-y_1}^{+y_1} (p_l - p_u)\, dy\, dx$$

$$= \frac{\rho_\infty}{2} V_\infty^2\, 2\,\pi\,\alpha\, \frac{d\,(y_1^2)}{dx} dx$$

This gives the lift as

$$L = \int_0^c dL$$

$$= \frac{\pi}{2} \frac{\rho_\infty}{2} V_\infty^2\, \alpha\, (2s)^2$$

This means that, whatever the planform is, the lift generated is the same if the span is the same. This can be so, because the dL depends on the gradient of planform only and if the initial gradient is small for a wing, it has to move later on in order to get the same $2s$. The lift coefficient becomes

$$C_L = \frac{L}{\frac{1}{2}\rho_\infty V_\infty^2 S}$$

where S is the planform area of the wing. The C_L can be expressed as

$$\boxed{C_L = \frac{\pi}{2} \mathbf{A}\, \alpha} \tag{1.22}$$

where **A** is the aspect ratio.

The general solution for flow past a slender body can be expressed as

$$\phi(x,y,z) = \phi_0(x,y,z) + \phi_\alpha(x,y,z) + g(x, M_\infty)$$

where for $\Delta\phi_0 = 0$,

$$v_n = \frac{V_\infty \dfrac{\partial z_b}{\partial x}}{\sqrt{1 - \left(\dfrac{\partial z_b}{\partial x}\right)^2}}$$

for $\Delta\phi_\alpha = 0$,

$$v_n = 0$$

and

$$\Delta \equiv \nabla^2$$

where Δ means "the change in" in mathematics. For example, if the variable "ϕ" stands for the potential function of the flow past an object, then "$\Delta\phi$" means "the change in the potential function" and ∇ represents the "total flux change = (field change in x direction) + (field change in y direction) + (field change in z direction)."

Example 1.9 A slender body of aspect ratio 11 cruises at sea level altitude at Mach 4 with an angle of attack of 3°. If the lift experienced by the body is 51.3 MN, determine the planform area of the body.

Solution

Given, $\mathbf{A} = 11$, $M = 4$, $\alpha = 3°$, $L = 51.3$ MN. At the sea level, $\rho = 1.225\,\text{kg/m}^3$, and $T = 288$ K.

The speed of sound is

$$\begin{aligned} a &= \sqrt{\gamma RT} \\ &= \sqrt{1.4 \times 287 \times 288} \\ &= 340\,\text{m/s} \end{aligned}$$

The flight speed is

$$\begin{aligned} V &= M\,a \\ &= 4 \times 340 \\ &= 1360\,\text{m/s} \end{aligned}$$

By Eq. (1.22), the lift coefficient is

$$C_L = \frac{\pi}{2}\mathbf{A}\,\alpha$$
$$= \frac{\pi}{2} \times 11 \times \frac{3}{57.3}$$
$$= 0.905$$

The lift is given by

$$L = \frac{1}{2}\rho V^2 S C_L$$

where S is the planform area. Therefore,

$$S = \frac{L}{\frac{1}{2}\rho V^2 C_L}$$
$$= \frac{51.3 \times 10^6}{\frac{1}{2} \times 1.225 \times 1360^2 \times 0.905}$$
$$= \boxed{50.04\,\text{m}^2}$$

1.6.1 Complex Analytic (Holomorphic) Function

Let us consider a wing without camber and without thickness, as shown in Fig. 1.14.

For this wing, the sectional lift coefficient is

$$dL = 2\pi q_\infty\, d(\alpha\, y_1^2)\, dx$$

where $\alpha = \alpha(x)$ and $q_\infty = \frac{1}{2}\rho_\infty V_\infty^2$.

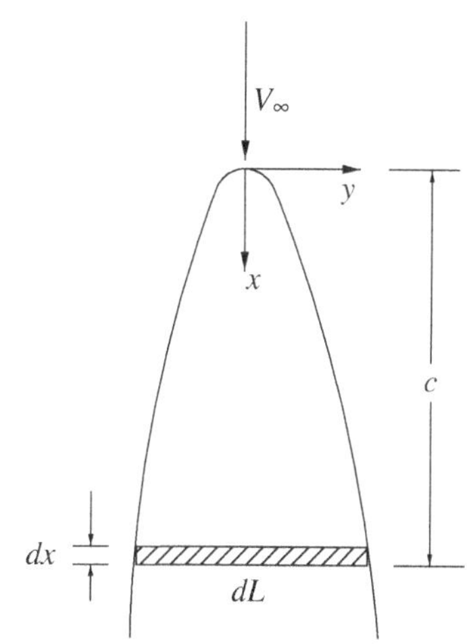

Figure 1.14 A slender wing body with zero camber and zero thickness.

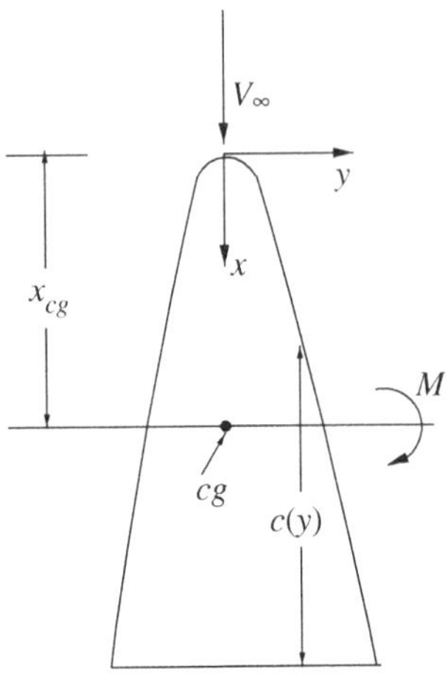

Figure 1.15 Pitching moment, M, of a slender wing.

Kutta condition (i.e. the flows from the upper and the lower portions of an airfoil rejoin at the trailing edge with no sudden change in the flow direction) is satisfied only for $dy_1/dx = 0$, at $x = c$. Kutta condition is fully satisfied for ogival and gothic wings that are more commonly used than the delta-shaped wings. For this case,

$$C_L = \frac{\pi}{2}\mathbf{A}\,\alpha$$

1.6.2 Pitching Moment About a Reference Axis (Axis Through *cg*)

The pitching moment, M, for the slender body as shown in Fig. 1.15 is

$$M = -\int_0^c (x - x_{cg})\,dL$$

Pitching moment coefficient becomes

$$C_M = \frac{M}{q_\infty S C_\mu}$$
$$= \frac{-2\pi\alpha}{SC_\mu}\left[(c - x_{cg})S^2 - \int_0^c y_1^2 dx\right]$$

where

$$C_\mu = \frac{1}{S}\int_{-s}^{+s} c^2 dy$$

This C_M cannot be expressed by a general relation.

For the slender body shown in Fig. 1.15,

$$C_\mu = \frac{2}{3}c$$

Therefore, the pitching moment coefficient becomes

$$\boxed{C_M = -\frac{3}{4}\pi\,\mathbf{A}\,\alpha\left(\frac{2}{3} - \frac{x_{cg}}{c}\right)} \tag{1.23}$$

From this relation, it is seen that, at $\frac{x_{cg}}{c} = \frac{2}{3}$, the pitching moment coefficient $C_M = 0$. This implies that the aerodynamic center (ac)[7] coincides with the center of pressure.

Example 1.10 For a slender wing of aspect ratio 4, flying with the center of gravity at 30% of the chord from the leading edge, if the pitching moment coefficient is −0.12, determine the angle of attack.

Solution

Given, $\mathbf{A} = 4$, $x_{cg} = 0.3\,c$, and $C_M = -0.12$.

By Eq. (1.23),

$$C_M = -\frac{3}{4}\pi\,\mathbf{A}\,\alpha\left(\frac{2}{3} - \frac{x_{cg}}{c}\right)$$

Therefore,

$$\begin{aligned}
-0.12 &= -0.75 \times \pi \times 4 \times \alpha\left(\frac{2}{3} - 0.3\right)\\
-0.12 &= -3.456\,\alpha\\
\alpha &= 0.12/3.456\\
&= 0.0347 \text{ radian}\\
&= 0.0347 \times 57.3\\
&= \boxed{2^\circ}
\end{aligned}$$

1.7 Stability Derivatives for Delta Wings

Let us consider the slender delta wing, as shown in Fig. 1.11, of aspect ratio $\mathbf{A} < 1$. For the given body, inner chord is c_i, outer chord is c_o. Let $\lambda = c_o/c_i$. The aspect ratio is

$$\mathbf{A} = \frac{b^2}{F} = \frac{2b}{c_i\,(1+\lambda)}$$

7 The location where the aerodynamic moment remains constant, that is the location where the aerodynamic force applied eliminates the problem of the movement of the center of pressure with angle of attack.

Let

$$F = \frac{1}{2}\, b\, c_i\, (1 + \lambda)$$

$$c_\mu = \frac{2}{3}\, c_i\, \frac{1 + \lambda + \lambda^2}{1 + \lambda}$$

In terms of c_μ and F, the lift and pitching moment coefficient are

$$C_L = \frac{L}{\frac{1}{2}\rho V_\infty^2 F} \tag{1.24}$$

$$C_M = \frac{M}{\frac{1}{2}\rho V_\infty^2 F c_\mu} \tag{1.25}$$

Also,

$$\frac{x_{\mathrm{ac}}}{c_i} = \frac{2}{3}\,(1 - \lambda)$$

$$\Omega_y = \frac{\omega_y\, c_\mu}{V_\infty}$$

The stability derivatives can be expresses as

$$\begin{aligned}
C_{L\alpha} &= \frac{\partial C_L}{\partial \alpha} \\
&= \frac{\pi}{2}\mathbf{A} \\
C_{M\alpha} &= \frac{\partial C_M}{\partial \alpha} \\
&= -\frac{3}{4}\pi\mathbf{A}\, \frac{1 + \lambda}{1 + \lambda + \lambda^2} \left[\frac{2}{3}(1 - \lambda) - \frac{x_{\mathrm{cg}}}{c_i}\right] \\
C_{Ly} &= \frac{\partial C_L}{\partial \Omega_y} \\
&= \frac{3}{4}\pi\mathbf{A}\, \frac{(1 + \lambda)^2}{(1 + \lambda + \lambda^2)^2} \left(1 - \frac{x_{\mathrm{cg}}}{c_i}\right) \\
C_{My} &= \frac{\partial C_M}{\partial \Omega_y} \\
&= -\frac{3}{8}\pi\mathbf{A}\, \frac{(1 + \lambda)^2}{(1 + \lambda + \lambda^2)^2} \left[3\left(\frac{x_{\mathrm{cg}}}{c_i}\right)^2 + (2\lambda - 5)\frac{x_{\mathrm{cg}}}{c_i} + \frac{9}{4} - \frac{3}{2}\lambda + \frac{3}{4}\lambda^2\right] \\
C_{L\dot{\alpha}} &= \frac{\partial C_L}{\partial \left(\dfrac{\alpha \cdot C_\mu}{V_\infty}\right)}
\end{aligned}$$

$$= \frac{1}{4}\pi\mathbf{A}\frac{(1+\lambda)(1+2\lambda)}{1+\lambda+\lambda^2}$$

$$C_{M\dot{\alpha}} = \frac{\partial C_M}{\partial\left(\frac{\alpha \cdot C_\mu}{V_\infty}\right)}$$

$$= \frac{3}{8}\pi\mathbf{A}\frac{(1+\lambda)^2}{(1+\lambda+\lambda^2)^2}\left[(1+2\lambda)\frac{x_{cg}}{c_i} - \left(\frac{3}{4}+\frac{3}{2}\lambda-\frac{3}{4}\lambda^2\right)\right]$$

$$C_y + C_{L\dot{\alpha}} = \frac{3}{4}\pi\mathbf{A}\frac{1+\lambda}{1+\lambda+\lambda^2}\left(\frac{4+2\lambda}{3} - \frac{x_{cg}}{c_i}\right)$$

$$C_{My} + C_{M\dot{\alpha}} = -\frac{9}{8}\pi\mathbf{A}\frac{(1+\lambda)^2}{(1+\lambda+\lambda^2)^2}\left(1-\frac{x_{cg}}{c_i}\right)^2$$

where c_i is the inner chord, and, for delta-wing, $\lambda \to 0$.

Example 1.11 A slender body (similar to that in Fig. 1.11), having inner chord 10 m, outer chord 10 cm, base 1 m, center of gravity at 30% of inner chord, flies at Mach 8, at the sea level. If the lift coefficient is 0.2, find the lift force and pitching moment coefficient.

Solution

Given, $c_i = 10$ m, $c_0 = 0.1$ m, $b = 1$ m, $x_{cg} = 0.3c_i$, $M_\infty = 8$, and $C_L = 0.2$.

At the sea level, $\rho_\infty = 1.225$ kg/m^3, $T = 288$ K.

The speed of sound, a_∞, and flight speed, V_∞, are

$$a_\infty = \sqrt{\gamma R T}$$
$$= \sqrt{1.4 \times 287 \times 288}$$
$$= 340 \text{ m/s}$$
$$V_\infty = M_\infty a_\infty$$
$$= 8 \times 340$$
$$= 2720 \text{ m/s}$$

Also,

$$\lambda = c_0/c_i$$
$$= 0.1/10 = 0.01$$
$$F = \frac{1}{2} b c_i (1+\lambda)$$
$$= \frac{1}{2} \times 1 \times 10 \times (1+0.01)$$
$$= 5.05$$

$$
\begin{aligned}
c_\mu &= \frac{2}{3}c_i\frac{1+\lambda+\lambda^2}{1+\lambda} \\
&= \frac{2}{3}\times 10\times\frac{1+0.01+0.01^2}{1+0.01} \\
&= \frac{2}{3}\times 10\times\frac{1.01}{1.01} \\
&= 6.667
\end{aligned}
$$

The lift becomes

$$
L = \frac{1}{2}\rho_\infty V_\infty^2 S C_L
$$

The area S is

$$
\begin{aligned}
S &= \frac{1}{2}\times b\times(c_i - c_0) + (b\times c_0) \\
&= \frac{1}{2}\times 1\times(10-0.01) + 1\times 0.01 \\
&= 0.5\times 1\times 9.99 + 0.01 \\
&= 4.995 + 0.01 \\
&= 5.005\,\mathrm{m}^2
\end{aligned}
$$

The lift is

$$
\begin{aligned}
L &= \frac{1}{2}\times 1.225\times 2720^2\times 5.005\times 0.2 \\
&= \boxed{453605.52\mathrm{N}}
\end{aligned}
$$

The pitching moment coefficient is

$$
\begin{aligned}
C_M &= \frac{M}{\frac{1}{2}\rho_\infty V_\infty^2 F c_\mu} \\
&= \frac{L\times x_{cg}}{\frac{1}{2}\times 1.225\times 2720^2\times 5.005\times 6.667} \\
&= \frac{453605.52\times 3}{\frac{1}{2}\times 1.225\times 2720^2\times 5.005\times 6.667} \\
&= \frac{1360816.56}{151209277.4} \\
&= \boxed{0.009}
\end{aligned}
$$

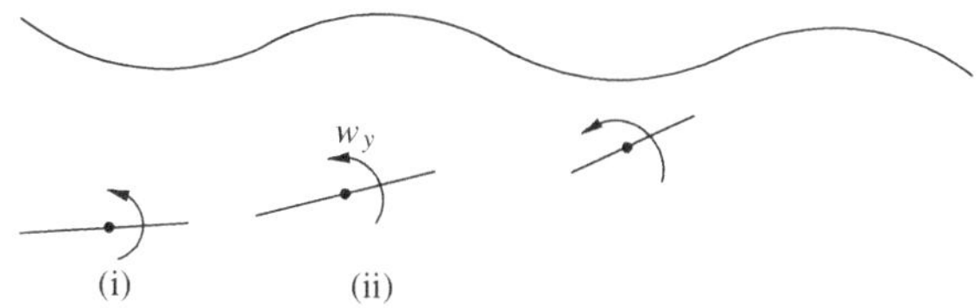

Figure 1.16 Flight path of a slender body.

1.7.1 Dynamic Stability Derivatives

Let us consider the wavy path, as shown in Fig. 1.16, followed by a slender body. The dynamic derivatives are necessary to study the dynamic stability. Such a flight of the wing can be studied as a superposition of the following two cases: (i) and (ii) marked in Fig. 1.16.

1. For constant w_y:
   The derivatives are C_{Ly} and C_{My}.
2. For constant $\dot{\alpha}$:
   The derivatives are $C_{L\dot{\alpha}}$ and $C_{M\dot{\alpha}}$ and

$$C_{M\alpha} = \frac{\partial C_M}{\partial \alpha}$$

Once the four derivatives $C_{L\alpha}$, $C_{M\alpha}$, $C_{L\dot{\alpha}}$, and $C_{M\dot{\alpha}}$ are known, then the stability (or the dynamic behavior) in known fully. These derivatives can be determined in the following manner.

Consider a slender body at an angle of incidence, α, to a freestream direction, flying with freestream velocity, V_∞, as shown in Fig. 1.17.

The angle of incidence can be expressed as

$$\alpha(x) = \frac{\omega_y}{V_\infty}(x - x_{cg})$$

This can be introduced into the sectional lift as

$$dL = 2\pi\, q_\infty\, d(\alpha\, y_1^2)$$

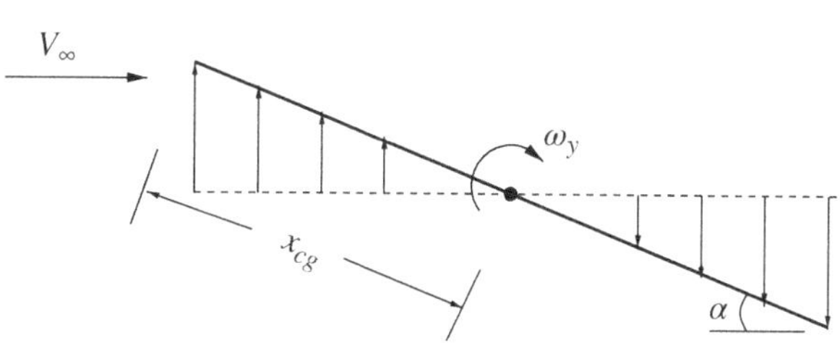

Figure 1.17 A slender body at an incidence.

to get the lift coefficient for delta-wing as

$$C_L = \frac{L}{q_\infty S} = 2\pi \frac{\omega_y c}{V_\infty} \frac{\mathbf{A}}{4} \left(1 - \frac{x_{cg}}{c}\right) \tag{1.26}$$

But the angular velocity is

$$\Omega = \frac{\omega_y C_\mu}{V_\infty} = \frac{2}{3} \frac{\omega_y c}{V_\infty}$$

Therefore,

$$\boxed{C_{Ly} = \frac{\partial C_L}{\partial \Omega_y} = \frac{3}{4}\pi \mathbf{A} \left(1 - \frac{x_{cg}}{c}\right)} \tag{1.26a}$$

Example 1.12 A flat plate of length 2 m and width 1 m flies at Mach 3 with incidence 3° at an altitude where the temperature is 200 K. If the center of gravity (cg) is at 0.5 m from the leading edge, determine the rotation about cg and the lift acting on the plate, assuming the density of air as 0.08 kg/m^3.

Solution

Given, $l = 2$ m, $c = 1$ m, $M = 3$, $\alpha = 3°$, $x_{cg} = 0.5$ m, $T = 200$ K, and $\rho = 0.08$ kg/m^3.

The speed of sound is

$$\begin{aligned} a &= \sqrt{\gamma R T} \\ &= \sqrt{1.4 \times 287 \times 200} \\ &= 283.5 \text{ m/s} \end{aligned}$$

The flight speed is

$$\begin{aligned} V &= M a \\ &= 3 \times 283.5 \\ &= 850.3 \text{ m/s} \end{aligned}$$

The aspect ratio of the plate is

$$\mathbf{A} = \frac{l}{c} = \frac{2}{1} = 2$$

The angle of attack in terms of rotation ω is

$$\alpha = \frac{\omega_y}{V}(x - x_{cg})$$

where y is the transverse axis. Therefore,

$$\begin{aligned} \omega_y &= \frac{\alpha V}{x - x_{cg}} \\ &= \frac{(3/57.3) \times 850.3}{c - 0.5c} \end{aligned}$$

$$= \frac{(3/57.3) \times 850.3}{0.5}$$
$$= 89 \text{ rev/s}$$

The lift coefficient, by Eq. (1.26), is

$$C_L = 2\pi \frac{\omega_y c}{V} \frac{\mathbf{A}}{4} \left(1 - \frac{x_{cg}}{c}\right)$$
$$= \frac{2\pi \times 89 \times 1}{850.5} \times \frac{2}{4} \times (1 - 0.5)$$
$$= \frac{89\pi}{850.5} \times 0.5$$
$$= 0.164$$

The lift is given by

$$L = \frac{1}{2} \rho V^2 S C_L$$
$$= \frac{1}{2} \times 0.08 \times 850.3^2 \times (2 \times 1) \times 0.164$$
$$= \boxed{9490.4 \text{ N}}$$

The Parameters $C_{L\dot{\alpha}}$ and $C_{M\dot{\alpha}}$

These derivatives are unsteady derivatives and they can be calculated only with unsteady slender body theory.

1.7.2 Unsteady Slender Body Theory

The continuity equation for steady flow is

$$\phi_{yy} + \phi_{zz} = 0$$

For unsteady flow, the continuity equation is

$$\phi_{yy} + \phi_{zz} - \frac{1}{a_\infty^2} \phi_{tt} = 0$$

where a_∞ is the speed of sound upstream of the body. The solution for this equation is $\phi(x, y, z, t)$.

For steady flow, the pressure coefficient is

$$C_p = -\frac{2}{V_\infty} \left(\frac{\partial \phi}{\partial x}\right)$$

The pressure coefficient for unsteady flow is

$$C_p = -\frac{2}{V_\infty} \left(\frac{\partial \phi}{\partial x} + \frac{1}{V_\infty} \frac{\partial \phi}{\partial t}\right)$$

On the surface of the body:

For steady flow, the potential function is

$$\boxed{\phi(x,\, y,\, 0) = V_\infty\, \alpha\, \sqrt{y_1^2 - y^2}}$$

The pressure coefficient is

$$C_p = -\frac{2}{V_\infty}\left(\frac{\partial\phi}{\partial x}\right)$$

For unsteady flow, the potential function is

$$\boxed{\phi(x, y, z, t) = V_\infty\, \dot{\alpha}\, t\, \sqrt{y_1^2 - y^2}}$$

This itself is the required solution.

The pressure coefficient is

$$C_p = \mp\frac{2}{V_\infty}\left(V_\infty\, \dot{\alpha}\, t \frac{y_1}{\sqrt{y_1^2 - y^2}}\frac{dy_1}{dx} + \dot{\alpha}\, \sqrt{y_1^2 - y^2}\right)$$

At $t = 0$, the pressure coefficient is

$$C_p = \mp\frac{2}{V_\infty}\left(\dot{\alpha}\, \sqrt{y_1^2 - y^2}\right)$$

This is an elliptic distribution over the span for unsteady case.

The lift is given by

$$L = \int_0^c \int_{-s}^{+s} (p_l - p_u)\, dydx$$

The lift coefficient is

$$C_L = 2\pi \frac{\dot{\alpha}\, C_\mu}{V_\infty} \cdot \frac{1}{C_\mu\, S}\int_0^c y_1^2 dx$$

For Delta wing:

$$C_L = \left(\frac{\pi}{4}\mathbf{A}\right)\left(\frac{\dot{\alpha}\, C_\mu}{V_\infty}\right)$$

$$C_{L\dot{\alpha}} = \frac{\partial C_L}{\partial\left(\dfrac{\dot{\alpha}\, C_\mu}{V_\infty}\right)}$$

$$= \frac{\pi}{4}\mathbf{A}$$

This unsteady analysis is possible only in the slender body theory, but not so easy for other bodies, because Kutta condition is not satisfied for them. In general, the pressure coefficient may be expressed as

$$C_p = \pm\frac{2}{V_\infty}\dot{\alpha}\sqrt{y_1^2 - y^2} \neq 0\,, \quad \text{when} \quad y = 0$$

However, this method of conformal mapping is not good for computer programming.

Example 1.13 If the coefficient of pressure for a slender body flying steadily at velocity V is –0.22, find the velocity potential function.

Solution

Given, $C_p = -0.22$.

The pressure coefficient can be expressed as

$$C_p = -\frac{2}{V}\left(\frac{\partial \phi}{\partial x}\right)$$

where ϕ is the velocity function, V is flight speed, and x is the flight direction. Therefore,

$$-0.22 = -\frac{2}{V}\left(\frac{\partial \phi}{\partial x}\right)$$

$$\frac{\partial \phi}{\partial x} = \frac{0.22\,V}{2}$$

$$= 0.11\,V$$

$$\phi = \boxed{0.11\,Vx}$$

1.8 Method of Vortex (Source and Sink) Distribution

Consider the control volume for the flow past a slender body, as shown in Fig. 1.18, which is suitable for computation.

The governing equation is

$$\Delta \phi = 0$$

Because each elemental circulation, $d\Gamma$, is a solution of the original potential flow equation for ϕ, the solution for the entire field can be obtained by summing $d\Gamma$ over the control volume.

The velocity inside the boundary is zero. The difference between the outer and inner tangential velocities is equal to the vorticity distribution, γ,

$$v_{to} - v_{ti} = \gamma$$

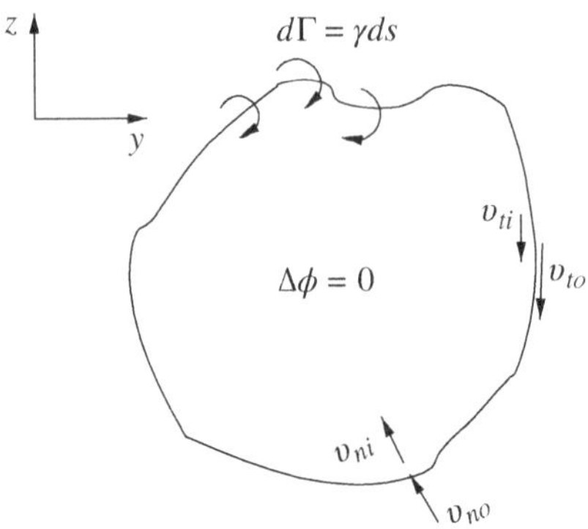

Figure 1.18 A control volume.

Similarly, the difference between the outer and inner normal velocities is equal to the source–sink distribution, q,

$$v_{no} - v_{ni} = q = \frac{\text{Source strength}}{\text{Length}}$$

If we consider vortex distribution, there is a jump in the tangential velocity and there is no normal velocity. This is exactly the solution required for $\Delta\phi_\alpha = 0$: the lift problem.

The given contour is

$$\frac{y_b}{y_1(x)} = \eta_b(\phi, x); \quad \frac{z_b}{y_1(x)} = \zeta_b(\phi, x); \quad 0 < \phi \leq 2\pi$$

For example, if the cross section is circular:

$$\eta_b = \cos\phi; \quad \zeta_b = \sin\phi$$

The unknown function: G_α

Referring to the difference with respect to ϕ and x,

$$G_\alpha = \frac{\gamma_\alpha}{V_\infty \alpha}\sqrt{\dot{\eta}_b^{\,2} + \dot{\zeta}_b^{\,2}}$$

where "$\dot{\eta}_b$" refers to differentiation with respect to ϕ and "$\dot{\zeta}_b$" refers to differentiation with respect to x.

Now we get the integral equation for the unknown function G_α as

$$\boxed{G_\alpha(\phi) - \frac{1}{2\pi}\int_0^{2\pi} k_1(\phi, \phi')\, G_\alpha(\phi')\, d\phi' = 2\dot{\zeta}_b} \tag{1.27}$$

Integral equation of the second kind, which is a singular integral equation, can be obtained as follows.

Consider the body in a vertical freestream flow field, as shown in Fig. 1.19.

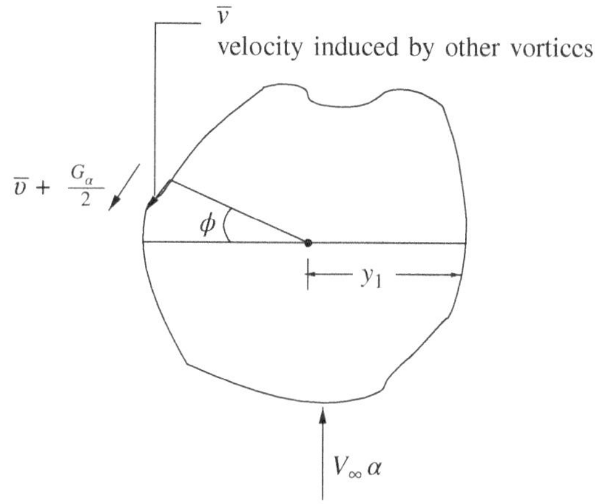

Figure 1.19 A slender wing in a vertical freestream.

The velocity inside the contour should be zero. Therefore,

$$\overline{v} - \frac{G_\alpha}{2} = 0$$

and

$$\overline{v} = \frac{1}{2\pi}\int_0^{2\pi} k_1(\phi, \phi')\, G_\alpha(\phi')\, d\phi' + 2\dot{\zeta}_b$$

This is the solution for the lift problem: the transverse flow problem.

1.8.1 Longitudinal Flow Problem

This is a problem of source–sink distribution. For this problem, in general, the velocity inside will not be zero. Therefore, to make the flow velocity zero a vortex distribution has to be superposed, so that the resultant velocity inside becomes zero.

This does not alter the normal velocity because no normal velocity is induced by vortices.

$$v_n = q$$

$$\frac{q}{V_\infty \frac{dy_1}{dx}}\sqrt{\dot{\eta}_b{}^2 + \dot{\zeta}_b{}^2} = Q_0$$

$$\eta_b' \rightarrow \eta_b \text{ and } \zeta_b' \rightarrow \zeta_b$$

In this case, the equation is not singular. But, the right-hand side is to be integrated over a singular function of the Cauchy type.

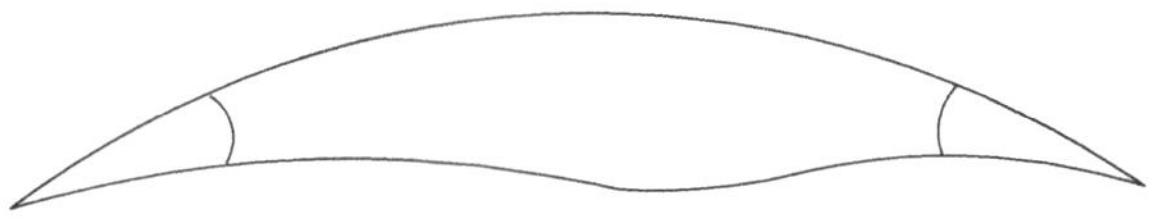

Figure 1.20 A slender cross section with corners.

1.8.2 For Cross Sections with Sharp Corners

For bodies with sharp corners such as that shown in Fig. 1.20 also, there is no problem in solving the flow field in principle, because only the potential function ϕ has to be chosen suitably and rest all the same as before.

1.8.3 General Solutions for Slender Configurations

Given, body geometry, as

$$y_1(x),\quad \frac{y_b}{y_1(x)} = \eta_b(\phi, x),\quad \frac{z_b}{y_1(x)} = \zeta_b(\phi, x)$$

where $0 \leq \phi \leq 2\pi$, $a(x)$ is the area distribution. Find the pressure coefficient C_p distribution on the body surface, and the forces acting on the body.

Transverse flow:

$$G_\alpha(\zeta) - \frac{1}{2\pi}\int_0^{2\pi} K_1(\zeta, \zeta')\, G_\alpha(\zeta')\, d\zeta' = 2\,\dot{\zeta}_b$$

$$K_1(\zeta, \zeta') = 2\,\frac{(\eta_b - \eta_b')\,\dot{\zeta}_b - (\zeta_b - \zeta_b')\,\dot{\eta}_b}{(\eta_b - \eta_b')^2 + (\zeta_b - \zeta_b')^2}$$

$$\phi_{ab}(\phi, x) = \left[\zeta_b + \frac{1}{2\pi}\int_0^{2\pi} G_\alpha(\zeta')\tan^{-1}\frac{\eta_b' - \eta_b}{\zeta_b - \zeta_b'}\, d\zeta'\right] V_\infty y_1\, \alpha$$

where superscript "dot" represents differentiation with respect to ζ, and "prime" is integration variable.

Longitudinal flow:

$$G_0(\zeta) - \frac{1}{2\pi}\int_0^{2\pi} K_1(\zeta, \zeta')\, G_0(\zeta')\, d\zeta' = \frac{1}{2\pi}\int_0^{2\pi} K_2(\zeta, \zeta')\, G_0(\zeta')\, d\zeta'$$

$$K_2(\zeta, \zeta') = 2\,\frac{(\eta_b - \eta_b')\,\dot{\zeta}_b - (\zeta_b - \zeta_b')\,\zeta_b}{(\eta_b - \eta_b')^2 + (\zeta_b - \zeta_b')^2}$$

$$Q_0 = \frac{\dfrac{\partial z_b}{\partial x}\sqrt{\dot{\zeta}_b^2 + \dot{\phi}_b^2}}{\dfrac{dy_1}{dx}\sqrt{1 + \left(\dfrac{\partial z_b}{\partial x}\right)^2}}$$

$$\phi_{0b}(\phi, x) = \frac{1}{2\pi}\left[\int_0^{2\pi} G_0(\zeta)\tan^{-1}\frac{(\eta_B - \eta_B')}{(\zeta_B - \zeta_B')}\, d\zeta' + \int_0^{2\pi} Q_0(\phi')\ln\frac{y_1}{c}\sqrt{(\eta_b - \eta_b')^2 + (\zeta_b - \dot{\zeta}_b)^2}\, d\zeta'\right] V_\infty y_1 \frac{dy_1}{dx}$$

Pressure coefficient:

$$C_p(\zeta, x, \alpha, M_\infty) = -2\frac{u_0}{V_\infty} - 2\frac{u_g(x, M_\infty)}{V_\infty} - \frac{G_0^2 + Q_0^2}{\dot{\eta}_b^2 + \dot{\zeta}_b^2}\left(\frac{dy_1}{dx}\right)^2$$

$$-2\left[\frac{u_\alpha}{\alpha V_\infty} + \frac{G_\alpha G_0}{\dot{\eta}_b^2 + \dot{\zeta}_b^2}\frac{dy_1}{dx}\right]\alpha$$

$$\frac{u_\alpha}{\alpha V_\infty} = \frac{1}{V_\infty}\left(\frac{\partial \phi_{\alpha b}}{\partial x}\right)_\zeta - G_\alpha\left[\frac{\eta_b\dot{\eta}_b + \phi_b\dot{\phi}_b}{\dot{\eta}_b{}^2 + \dot{\zeta}_b{}^2}\frac{dy_1}{dx} + \frac{\dot{\eta}_b\left(\frac{\partial \eta_b}{\partial x}\right)_\zeta + \dot{\zeta}_b\left(\frac{\partial \zeta_b}{\partial x}\right)_\zeta}{\dot{\eta}_b{}^2 + \dot{\zeta}_b{}^2}y_1\right]$$

$$\frac{u_0}{\alpha V_\infty} = \frac{1}{V_\infty}\left(\frac{\partial \Phi_{\alpha b}}{\partial x}\right)_\zeta - G_0\left[\frac{\eta_b\dot{\eta}_b + \phi_b\dot{\zeta}_b}{\dot{\zeta}_b^2 + \dot{\zeta}_b^2}\frac{dy_1}{dx}\right.$$

$$\left.+\frac{\dot{\eta}_b\left(\frac{\partial \eta_b}{\partial x}\right)_\zeta - \dot{\zeta}_b\left(\frac{\partial \zeta_B}{\partial x}\right)_\zeta}{\dot{\eta}_b^2 + \dot{\zeta}_b^2}y_1\right]\frac{dy_1}{dx}$$

$$-Q_0\left[\frac{\eta_b\dot{\zeta}_b - \zeta_b\dot{\eta}_b}{\dot{\eta}_b{}^2 + \dot{\zeta}_b^2}\frac{dy_1}{dx} + \frac{\dot{\zeta}_b\left(\frac{\partial y_b}{dx}\right)_\phi - \dot{\eta}_b\left(\frac{\partial \zeta}{\partial x}\right)_\zeta}{\dot{\eta}_b^2 + \dot{\zeta}_b^2}y_1\right]\frac{dy_1}{dx}$$

$u_g = \frac{dg}{dx}$, $M_\infty < 1$:

$$g = \frac{V_\infty c}{4\pi}\left[\frac{2}{c}\frac{da}{dx}\ln\frac{\sqrt{1-M_\infty^2}}{2} - \frac{1}{c}\left(\frac{dA}{dx}\right)_{\xi=1}\ln(1-\xi)\right.$$

$$\left.-\int_0^\xi \frac{d^2a}{dx^2}\ln(\xi' - \xi)\,d\xi' + \int_\xi^1 \frac{d^2a}{dx^2}\ln(\xi' - \xi)\,d\xi'\right]$$

$\xi = \frac{x}{c}$, $M_\infty > 1$:

$$g = \frac{V_\infty c}{2\pi}\left[\frac{1}{c}\frac{da}{dx}\ln\frac{\sqrt{1-M_\infty^2}}{2} - \int_0^\xi \frac{d^2a}{dx^2}\ln(\xi' - \xi)\,d\xi'\right]$$

$$C_N = \frac{N}{q_\infty S} = \frac{c^2}{S}\int_0^1 \frac{y_1}{c}\int_0^{2\pi} C_p\,\dot{\eta}_b\,d\zeta\,d\xi$$

$$C_T = \frac{T}{q_\infty S} = \frac{c^2}{S}\int_0^1 \frac{dy_1}{dx}\frac{y_1}{c}\int_0^{2\pi} C_p\,(\eta_b\,\dot{\zeta}_b - \zeta_b\,\dot{\eta}_b)\,d\zeta\,d\xi$$

$$C_L = C_N - C_T\alpha \approx C_N$$

$$C_D = C_T + C_N\alpha$$

$$C_M = \frac{M}{q_\infty S c_\mu} = \frac{c^2}{S}\frac{c}{c_\mu}\int_0^1 (\xi - \xi_{cg})\frac{y_1}{c}\int_0^{2\pi} C_p \dot{\eta}_b \, d\phi \, d\xi$$

The first term in the C_p expression above is for the problem with $\alpha = 0$. The second term is for the space influence function. The third term is from the second integral equation. The fourth term (second line) is the coupling term.

In the relation,

$$\frac{u_\alpha}{\alpha V_\infty} = \frac{1}{V_\infty}\left(\frac{\partial \phi_{\alpha_b}}{\partial x}\right)_\zeta - G_\alpha\left[\frac{\eta_b\dot{\eta}_b + \zeta_b\dot{\zeta}_b}{\dot{\eta}_b{}^2 + \dot{\zeta}_b{}^2}\frac{dy_1}{dx} + \frac{\dot{\eta}_b\left(\frac{\partial \eta_b}{\partial x}\right)_\zeta + \dot{\zeta}_b\left(\frac{\partial \zeta_b}{\partial x}\right)_\zeta}{\dot{\eta}_b{}^2 + \dot{\zeta}_b{}^2} y_1\right]$$

the term

$$\left[\frac{\eta_b\dot{\eta}_b + \zeta_b\dot{\zeta}_b}{\dot{\eta}_b{}^2 + \dot{\zeta}_b{}^2}\frac{dy_1}{dx} + \frac{\dot{\zeta}_b\left(\frac{\partial y_b}{\partial x}\right)_\phi + \dot{\eta}_b\left(\frac{\partial \phi_b}{\partial x}\right)_\zeta}{\dot{\eta}_b{}^2 + \dot{\zeta}_b{}^2} y_1\right]$$

is the correction factor for ϕ_α being on the body and not along the x axis. This gives rise to the term in the square braces above.

1.8.4 Space Influence for $g(x, M_\infty)$

For the adjacent points on the surface of a slender body, as shown in Fig. 1.21, the difference in the correction factor gives rise to

$$\left[\frac{\eta_b\dot{\eta}_b + \phi_b\dot{\phi}_b}{\dot{\eta}_b{}^2 + \dot{\phi}_b{}^2}\frac{dy_1}{dx} + \frac{\dot{\phi}_b\left(\frac{\partial y_b}{\partial x}\right)_\phi + \dot{\eta}_b\left(\frac{\partial \phi_b}{\partial x}\right)_\phi}{\dot{\eta}_b{}^2 + \dot{\phi}_b{}^2} y_1\right]\frac{dy_1}{dx}$$

From $u_g = dg/dx$, when these are solved, the entire configuration will be fully known.

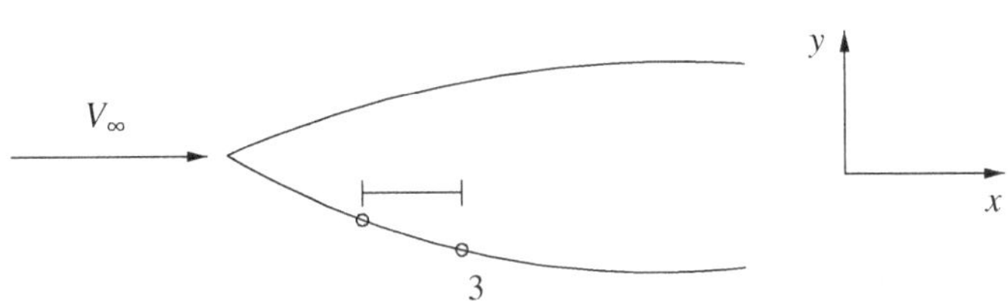

Figure 1.21 Two adjacent points on a slender wing.

For example, for body of revolution, the problem becomes a transverse flow problem. For this case

$$k_1 = 2\left[\frac{(\cos\phi - \cos\phi') - (\sin\phi - \sin\phi')\sin\phi}{(\cos\phi - \cos\phi')^2 + (\sin\phi - \sin\phi')^2}\right]$$

$$= 2\left[\frac{1 - \cos\phi'\cos\phi - \sin\phi'\sin\phi}{2 - 2\cos\phi'\cos\phi - 2\sin\phi'\sin\phi}\right]$$

$$= 1$$

When $k_1 \equiv 1$, this is a simple problem. The integral equation gives

$$G_\alpha - \frac{1}{2\pi}\int_0^{2\pi} G_\alpha(\phi')d\phi' = 2\cos\phi$$

But

$$G_\alpha = 2\cos\phi = \frac{\gamma_\alpha}{V_\infty\alpha} = \frac{V_t}{V_\infty\alpha}$$

because

$$\frac{\gamma_\alpha}{V_\infty\alpha}\sqrt{\dot{\eta}_b^{\,2} + \dot{\zeta}_b^{\,2}} = G_\alpha$$

The result

$$2\cos\phi = \frac{V_t}{V_\infty\alpha}$$

is the classical problem of two-dimensional flow around a circular cylinder.

Longitudinal flow:

For this flow, $Q_0 = 1$, and

$$G_0(\phi) - \frac{1}{2\pi}\int_0^{2\pi} G_\alpha(\phi')d\phi' = \frac{1}{2\pi}\int_0^{2\pi}\cot\frac{\phi - \phi'}{2}\,d\phi'$$

The velocity components at any point on the body will be as shown in Fig. 1.22. We have

$$\frac{q}{V_\infty} = \frac{dy_1}{dx}$$

This gives $Q_0 = 1$. The right-hand side of the above equation is the integral equation

$$\int_0^{2\pi}\cot\frac{\phi - \phi'}{2}\,d\phi' = 0$$

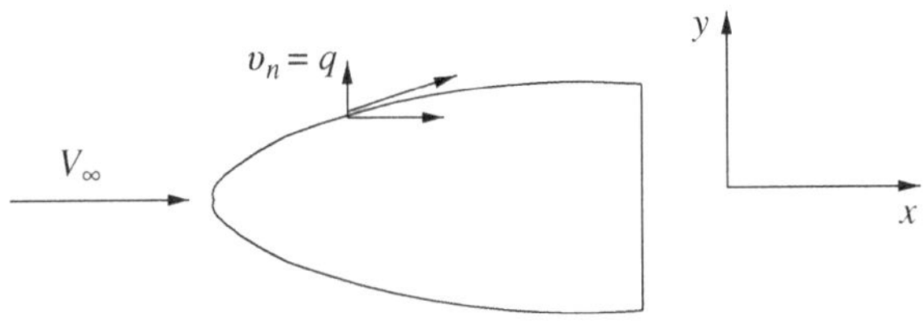

Figure 1.22 Velocity components at a point on the surface of a slender body.

Therefore,

$$G_0(\phi) = 0$$

Also,

$$\Phi_{ab}(\phi, x) = 2V_\infty \,\alpha\, y_1 \sin\phi$$

$$\Phi_{0b}(\phi, x) = V_\infty y_1 \frac{dy_1}{dx} \ln\left(\frac{y_1}{c}\right)$$

The relation

$$\frac{u_\alpha}{V_\infty \alpha} = 2\sin\phi \frac{dy_1}{dx}$$

implies that in the relation

$$\frac{u_\alpha}{\alpha V_\infty} = \frac{1}{V_\infty}\left(\frac{\partial \phi_{\alpha_b}}{\partial x}\right)_\phi - G_\alpha \left[\frac{\eta_b \dot{\eta}_b + \zeta_b \dot{\zeta}_b}{\dot{\eta}_b^{\,2} + \dot{\zeta}_b^{\,2}} \frac{dy_1}{dx} + \frac{\dot{\eta}_b \left(\frac{\partial \eta_b}{\partial x}\right)_\phi + \dot{\phi}_b \left(\frac{\partial \phi_b}{\partial x}\right)_\phi}{\dot{\zeta}_b + \dot{\zeta}_b^{\,2}} y_1 \right] \frac{dy_1}{dx}$$

$$-Q_0 \left[\frac{\eta_b \dot{\phi}_b - \phi_b \dot{\eta}_b}{\dot{\eta}_b^{\,2} + \dot{\phi}_b^{\,2}} \frac{dy_1}{dx} + \frac{\dot{\phi}_b \left(\frac{\partial y_b}{\partial x}\right)_\phi - \dot{\zeta}_b \left(\frac{\partial \phi_b}{\partial x}\right)_\phi}{\dot{\zeta}_b^{\,2} + \dot{\phi}_b^{\,2}} y_1 \right] \frac{dy_1}{dx} = 0$$

and

$$\frac{u_0}{V_\infty \alpha} = \frac{d^2}{dx^2}\left(\frac{y_1^2}{2}\right) \ln\left(\frac{y_1^2}{c}\right) + \left(\frac{dy_1}{dx}\right)^2 - \left(\frac{dy_1}{dx}\right)^2$$

implies that

$$\left[\frac{\eta_B \dot{\eta}_B + \zeta_B \dot{\zeta}_B}{\dot{\eta}_B^{\,2} + \dot{\zeta}_B^{\,2}} \frac{dy_1}{dx} + \frac{\dot{\eta}_B \left(\frac{\partial \eta_B}{\partial x}\right)_\phi + \dot{\zeta}_B}{\dot{\zeta}_B + \dot{\zeta}_B^{\,2}} y_1 \right] = 0$$

because

$$\eta_b \cos\phi \neq \eta_b(x)$$

and

$$\zeta_b = \sin\phi \neq \zeta_b(x)$$

The pressure due to angle of attack α is

$$\begin{aligned} C_p - C_{p\alpha} &= -2\frac{u}{V_{\infty x}}\,\alpha \\ &= -4\frac{dy_1}{dx}\sin\phi\,\alpha \end{aligned}$$

This is proportional to $\frac{dy_1}{dx}$, and $\sin\phi$. At any arbitrary section of the body, $\left(\frac{dy_1}{dx} = \text{constant}\right)$ the pressure coefficient is proportional to $\sin\phi$. That is, $(C_p - C_{p\alpha})$ is negative (–ve) on the upper surface and positive (+ve) on the lower surface.

For the cross section of a slender body, as shown in Fig. 1.21,

$$\begin{aligned} C_N &= \frac{c^2}{S}\int_0^1 \frac{y_1}{c}\int_0^{2\pi} C_p\,\dot{\eta}_b\,d\phi\,d\xi \\ &= \frac{4\pi\,\alpha}{S}\int_0^c d\left(\frac{y_1^2}{2}\right) \end{aligned}$$

For the slender body as shown in Fig. 1.22, we have

$$C_N = \frac{\pi}{2}\mathbf{A}\,\alpha$$

and for the slender body as shown in Fig. 1.23,

$$C_N = 0$$

For this case, the views showing the surface area and flow direction, respectively, are shown in Figs. 1.24 and 1.25.

For a slender body in an inviscid flow, to have a finite C_N (i.e. to have $C_N \neq 0$), there should be a base span. For such bodies, C_N will be the same for any planform. The moment coefficient is given by

$$C_M = 4\frac{1}{Sc_\mu}\pi\,\alpha\int_0^c (\alpha - x_{cg})\,d\left(\frac{y_1^2}{2}\right)$$

The integration in this relation is partial integration.

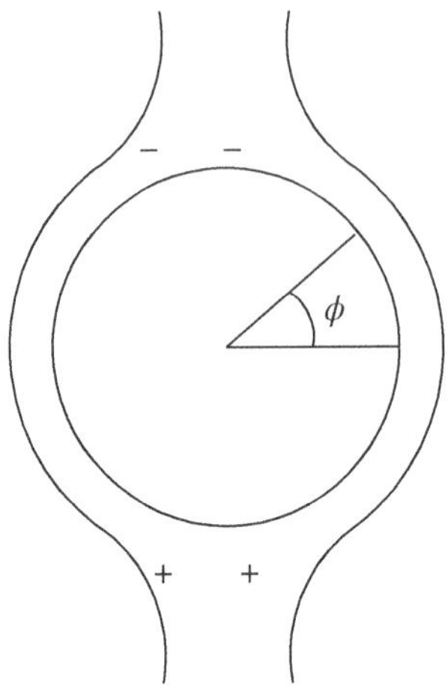

Figure 1.23 Cross section of a slender body.

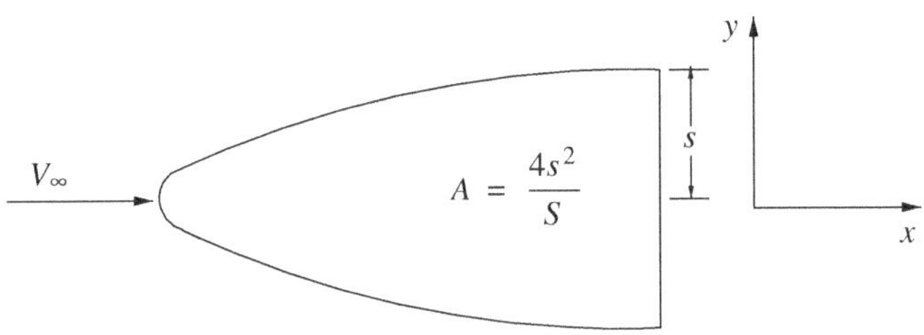

Figure 1.24 Surface area of a slender body.

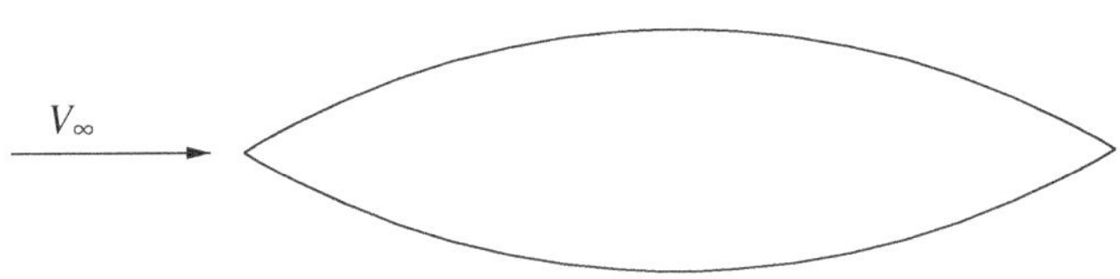

Figure 1.25 A slender body in an uniform flow.

Considering a slender body of volume **V**, as shown in Fig. 1.22, the C_M can be expressed as

$$C_M = 4\frac{1}{Sc_\mu}\pi\,\alpha\int_0^c \frac{y_1^2}{2}\,dx$$
$$= 2\frac{\mathbf{V}}{Sc_\mu}\,\alpha$$

Another way of nondimensionalizing the moment, M, is

$$\boxed{\frac{M}{q_\infty \mathbf{V}} = 2\,\alpha} \tag{1.28}$$

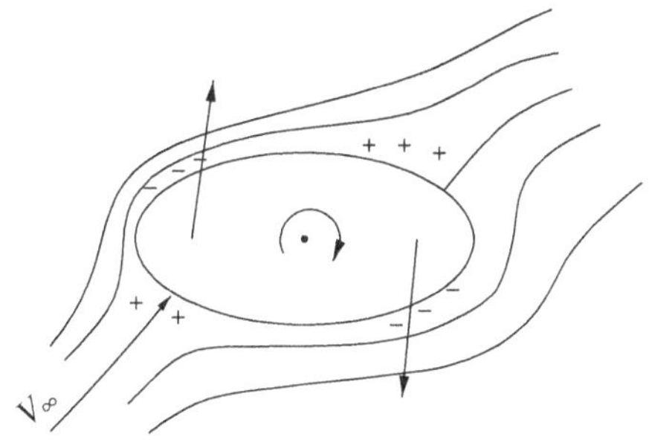

Figure 1.26 Flow at an angle, past a slender body.

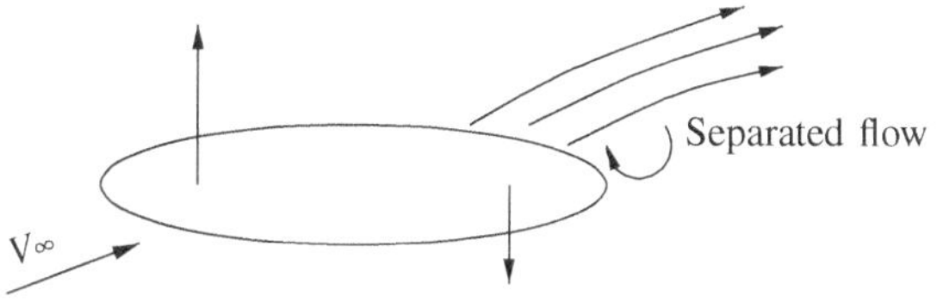

Figure 1.27 Flow separation due to large angle of attack.

This is a constant and independent of axis.

For the body in a flow field, as in Fig. 1.26, the normal force coefficient is zero, that is

$$C_N = 0$$

and the pitching moment coefficient is

$$\begin{aligned} C_M &= \frac{M}{q_\infty \mathbf{V}} \\ &= \text{constant} \end{aligned}$$

That is, C_M is independent of axis. This is a destabilizing moment.

It is seen that these bodies (illustrated in Figs. 1.22 and 1.26) are good in stability and the drag is also low.

However, they are bad in yaw, that is, any side wind can tilt the body making the angle of attack, α, large, leading to flow separation, as shown in Fig. 1.27, thus destroying the pressure there.

1.9 Cone in Supersonic Flow

Let us consider an axisymmetric cone in a supersonic flow, as shown in Fig. 1.28. This is an axisymmetric flow.

For this case,

$$\frac{dy_i}{dx_i} = \tan w_0 = \text{constant}$$

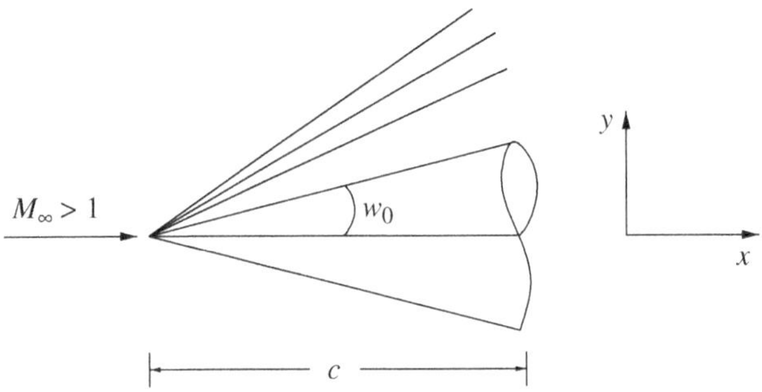

Figure 1.28 An axisymmetric cone in a supersonic flow.

$$\begin{aligned} A(x_i) &= \pi y_1^2 \\ &= \pi x_1^2 \tan^2 w_0 \\ y_1 &= x_1 \tan w_0 \end{aligned}$$

For $M_\infty > 1$, the normal component of velocity (u_g) at any point on the body contour is

$$u_g = \frac{dg}{dx} = V_\infty \tan^2 w_0 \left[\ln \frac{\sqrt{M_\infty^2 - 1}}{2} - \ln \frac{x}{c} \right]$$

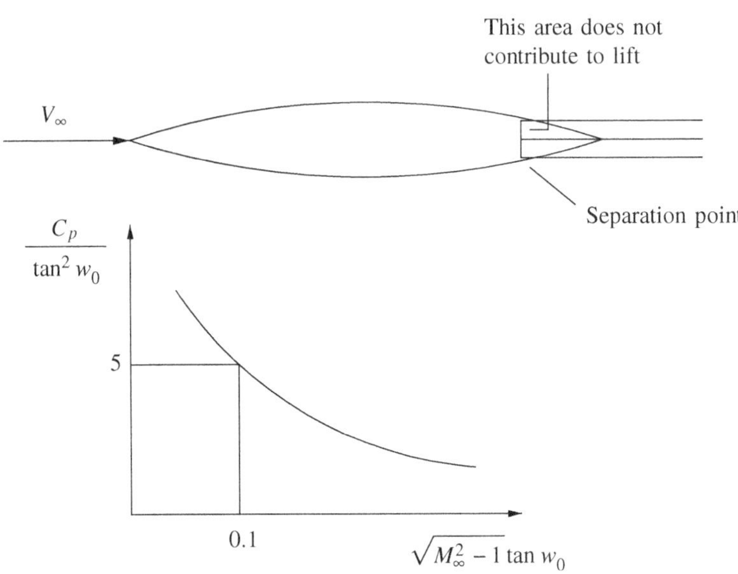

Figure 1.29 Flow separation due to large angle of attack.

Now the total C_p can be obtained by adding the following to $C_p - C_{p_{\alpha=0}}$, derived earlier.

$$C_{p_{\alpha=0}} = 2\frac{u_0}{V_\infty} - 2\frac{u_g}{V_\infty} - \left(\frac{dy_1}{dx}\right)^2 \; : \; \frac{G_0^2 + Q_0^2}{\dot{\eta}_b^{\,2} + \dot{\zeta}_b^{\,2}} = 1$$

$$C_{p_{\alpha=0}} = \tan^2 w_0 \left[2 \ln 2 - 1 - 2\ln\left(\sqrt{M_\infty^2 - 1}\tan w_0\right)\right]$$

This is the conical problem, as illustrated in Fig. 1.29. The pressure coefficient $C_{p_{\alpha=0}}$ does not depend on the freestream Mach number M_∞ as such but depends on $\sqrt{M_\infty^2 - 1}\tan w_0$, as seen from the expression. That is, all cones having this factor same will have the same $C_{p_{\alpha=0}}$. This is also the classical result of Gothert's and Prandtl and Glauert similarity rules.

1.10 Optimization (Supersonic Flow)

Given the volume of the body, what is the geometry which gives the minimum drag, or to find a body which gives the minimum drag for a given situation. For example, minimum drag for a given lift. For this kind of problem, the space influence function is simple and the expression for drag is available in an explicit simple form.

1.10.1 Zero-Lift Case ($\alpha = 0$)

The expression for the thrust coefficient, C_T, is very complicated to evaluate.

Consider a slender body in an inviscid flow, as shown in Fig. 1.30. For this body. the ratio of thrust T to freestream dynamic pressure q_∞ is given by

$$\frac{T}{q_\infty} = -\frac{1}{2\pi}\int_0^c\int_0^c A''(x)A''(x_1)\ln|x - x_1|\,dx\,dx_1 + \frac{1}{2\pi}A'(c)\int_0^c A''(x_1)\ln|c - x_1|\,dx_1 - \frac{1}{V_\infty}\oint_{x=c}(\Phi_0 + g)\frac{\partial(\Phi_0 + g)}{\partial n}\,ds$$

Such simple form is not available for subsonic flow.

1. This expression is simple, because it involves only $A(x)$, the area distribution. So, instead of treating the actual body as such, we can treat it as a body of revolution with the same $A(x)$, leading to an equivalent body of revolution.
2. The equation does not involve the freestream Mach number, M_∞, if it is a body of revolution.
3. The only place where the actual contour has to be considered is in the line integral, and the freestream Mach number, M_∞, enters here.

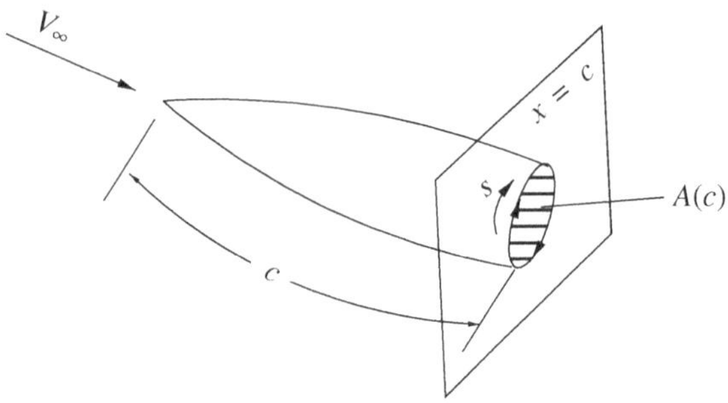

Figure 1.30 Slender body in a flow.

For a closed body of revolution:

$$\oint \phi_0 \frac{\partial \phi_0}{\partial n} ds = 0$$

This means that the thrust, T, is independent of M_∞.

We consider only wings. In other cases, the equations are difficult.

Wings:

Consider the wing as shown in Fig. 1.31. For this wing [5]

$$-\frac{1}{V_\infty} \oint \Phi_0 \frac{\partial \Phi_0}{\partial n} ds = c^2 \frac{\overline{a}'(1)}{2\pi} H_0(1)$$

$$\frac{x}{c} = \xi, \quad \frac{a}{c^2} = \overline{a},$$

$\overline{a}'$ is the differentiation with respect to ξ

$\beta = \sqrt{M_\infty^2 - 1}$ is the Prandtl factor

$$H_0(1) = \overline{k} - \ln\left(\beta \frac{s}{c}\right)$$

Example 1.14 Find the angle of attack for a body of volume **V**, length c, $\overline{a}_1 = a/c^2$, so that the drag is a minimum, that is, the thrust T required to overcome the drag is a minimum.

Solution

Given, volume = **V**, length is c, $\overline{a}_1$, find $a(x)$ so that T is minimum.

Let as assume

$$\overline{a}'' = \frac{2}{\sin\theta} \sum_{n=0}^{\infty} a_n \cos n\theta$$

The axial distance x varies from 0 to c; the nondimensional axial distance, ξ, varies from 0 to 1; θ varies from 0 to π.

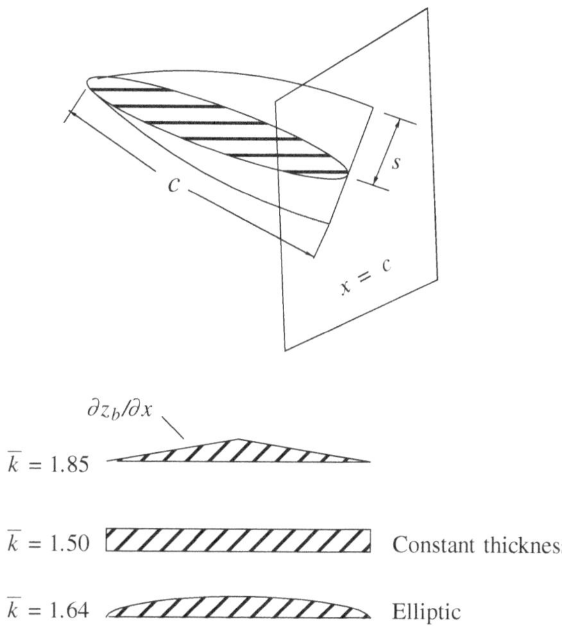

Figure 1.31 Slender wing of different cross-sectional shapes. Source: Adapted from Mangler and Smith [6].

The drag is $T(a_0, a_1, a_2, \ldots, a_n)$. This drag, which depends on all the n coefficients $A_0, A_1, A_2, \ldots, A_n$, is to be minimized. Therefore, the conditions to be satisfied are

$$\frac{\partial T}{\partial a_0} = 0$$

$$\frac{\partial T}{\partial a_1} = 0$$

$$\vdots$$

$$\frac{\partial T}{\partial a_n} = 0$$

This set of $(n + 1)$ can be easily solved for $a_0, a_1, a_2, \ldots, a_n$.

The nondimensionalized volume is

$$\overline{\mathbf{V}} = \frac{\mathbf{V}}{c^3}$$

The coefficients become

$$a_0 = \frac{8\mathbf{V}}{\pi}\,(x^* - 0.8)$$

$$a_1 = \frac{8\mathbf{V}}{\pi}\,(1.6 - x^*)$$

$$a_2 = \frac{8\mathbf{V}}{\pi}(1.2 - x^*)$$

where $\overline{a}(1) = x^* \times 2\overline{\mathbf{V}}$, that is, $x^* \neq 0$ only for bodies with base area $\overline{a}(1) \neq 0$ and

$$a_n = 0 \quad \text{for } n \geq 3$$

That is, only three of the Fourier series need to be taken.

Wings:
For this case, $x^* = 0$, and

$$\frac{T}{q_\infty c^2\overline{\mathbf{V}}^2} = \frac{128}{\pi}\left[1 - 0.16\left(2\ln 2 + 3 - \overline{\mathbf{V}} + \ln \beta\frac{s}{c}\right)\right]$$
$$= \frac{128\,k_0}{\pi}$$

where $k_0 = \left[1 - 0.16\left(2\ln 2 + 3 - \overline{\mathbf{V}} + \ln \beta\frac{s}{c}\right)\right]$

Body of revolution:
For this case, $x^* = 0$, and

$$\frac{T}{q_\infty c^2\overline{\mathbf{V}}^2} = \frac{128}{\pi}$$

This is independent of freestream Mach number M_∞. It is a closed body, that is, $\overline{a}(1) = 0$. This is called *Sears–Haack body.*[8]

The term k_0 in the above expression is the ratio of wing drag to the drag of Sears–Haack body for the same volume. k_0 could be made less than 1.

For the Sears–Haack body,

$$a_0 = 0; \quad a_1 = 0, \quad a_2 = \frac{32}{\pi}\overline{\mathbf{V}}.$$

$$\overline{a} = \frac{4}{\pi}\overline{\mathbf{V}}\left(\sin\theta - \frac{\sin 3\theta}{3}\right)$$

Often the drag of the body under consideration is compared to that of the Sears–Haack body.

If c is large, wave drag is small, but the friction drag will be more and vice versa. So the optimization could be carried out with respect to c so that the total drag is a minimum.

8 The Sears–Haack body is the shape with the lowest theoretical wave drag in supersonic flow, for a given body length and a given volume. The mathematical derivation assumes small-disturbance (linearized) supersonic flow, which is governed by the Prandtl–Glauert equation.

1.10.2 Optimization for Total Drag to Be a Minimum

Drag due to friction is

$$\frac{D_f}{q_\infty} \approx 2\,\overline{C}_f\, S$$

The total drag is

$$\begin{aligned}\frac{D}{q_\infty} &= \frac{T}{q_\infty} + \frac{D_f}{q_\infty} \\ &= \frac{128}{\pi}\, k_0 \frac{\mathbf{V}^2}{c^4} + 2\,\overline{C}_f \hat{k} s\, c\end{aligned}$$

where $\hat{k} = S/(c\,s)$. This depends on the planform shape. For optimum drag,

$$\frac{\partial (D/q_\infty)}{\partial c} = 0$$

which gives

$$\boxed{C_{D_{\text{opt}}} = \left[\frac{128}{\pi}\, k_0 \frac{\mathbf{V}^2}{2\,\overline{C}_f \hat{k}\frac{S}{c}}\right]^{1/6}} \tag{1.29}$$

k_0 depends on $A(x)$.

1.11 Lift-Case

For this case, for drag optimum, the drag due to lift; that is, the induced drag should be a minimum.

$$\frac{D_L}{q_\infty c^2} = \frac{2}{\pi} \int_{-1}^{+1} \int_{-1}^{+1} \frac{\gamma(\eta)}{\eta - \eta_1} \frac{\partial \gamma}{\partial \eta_1}\, d\eta\, d\eta_1 + [\cdots\cdot] - \{\cdots\cdot\} + (\cdots\cdot)$$

The induced drag is a minimum if $\gamma(\eta)$ is an elliptic distribution (i.e. elliptic distribution for the span-wise distribution of lift).

The second term is

$$[\cdots\cdot] = \frac{\beta^2 S^2}{8\,c^2} \left[-\frac{1}{2\,\pi} \int_{-1}^{1} \int_{-1}^{1} \Delta \hat{p}(1,\, \eta) \Delta \hat{C}_p(1,\, \eta_1)\, \ln{(\eta - \eta_1)}\, d\eta\, d\eta_1\right]$$

where $\Delta \hat{p}(1,\, \eta)$ is a base drag contribution. This base drag is zero for wings with parallel ends, that is, the Kutta condition is satisfied. For this, the base drag contribution term can be expressed as

$$\Delta \hat{C}_p = \frac{p_1 - p_u}{q_\infty\, S}$$

Hence, the optimum drag expression becomes

$$\frac{D_L}{q_\infty c^2} = \frac{2}{\pi}\int_{-1}^{+1}\int_{-1}^{+1}\frac{\gamma(\eta)}{\eta-\eta_1}\frac{\partial\gamma}{\partial\eta_1}\,d\eta\,d\eta_1$$

$$+\frac{\beta^2 S^2}{8c^2}\left[-\frac{1}{2\pi}\int_{-1}^{+1}\int_{-1}^{+1}\Delta\hat{p}(1,\eta)\Delta\hat{C}_p(1,\eta)\,\ln|\eta-\eta_1|\,d\eta\,d\eta_1\right.$$

$$-\frac{1}{2\pi}\int_0^1\int_0^1\sigma''(\xi)\sigma''(\xi_1)\,\ln|\xi-\xi_1|\,d\xi\,d\xi_1$$

$$\left.+\frac{\sigma'(1)}{\pi}\int_0^1\sigma''(\xi_1)\,\ln(1-\xi_1)\,d\xi_1+\frac{[\sigma'(1)]^2}{2\pi}\left(\frac{1}{2}-\ln\frac{\beta S}{2c}\right)\right]$$

the last three terms give the wave drag.

Example 1.15 For a wing with lift coefficient C_L, projected area S, and chord c, if the drag has to be minimum, find $\sigma(x)$, where $\sigma = 2\pi\left(\frac{y_1}{s}\right)^2\alpha(x)$.

Solution

Given, C_L, S/c, $\hat{k}$, and $\sigma = 2\pi\left(\frac{y_1}{s}\right)^2\alpha(x)$. Find $\sigma(x)$ for D_L minimum.

In the expression for $D_L/(q_\infty c^2)$, only the first two terms can be derived from the simple slender body theory. The last two terms are derived by Adams and Sears, which is something more than the simple slender body theory. The second term is zero for normal wings.

Nondimensional thickness for the wing is defined as

$$\overline{\tau_m} = \frac{\mathbf{V}}{Sc}$$

where $\mathbf{V}$ is the volume.

Example 1.16 Consider the Concorde wing flying at Mach 2.2 at 20 km altitude. For this altitude and Reynolds number, find the optimal gothic wing.

For this altitude and velocity let us find the Reynolds number. For the given wing, at the flight condition, it is reasonable to take the average skin friction coefficient as $\overline{C_f} = 1.35 \times 10^{-5}$, the lift as $L = 9.8$ MN and the volume as $\mathbf{V} = 400\,\text{m}^3$.

Therefore,

$$\overline{\tau_m} = \frac{V}{Sc} = 0.016$$

Further, x_{cg}/c also can be specified.

Solution

For the given problem, $\beta = \sqrt{M_\infty^2 - 1} \approx 2$. At 20 km altitude, the freestream flow density and sound speed are $\rho_\infty = 0.08816\,\mathrm{kg/m^3}$ and $a_\infty = 295\,\mathrm{m/s}$. Therefore,

$$V_\infty = M_\infty a_\infty$$
$$= 2.2 \times 295$$
$$= 649\,\mathrm{m/s}$$
$$q_\infty = \frac{1}{2}\rho_\infty V_\infty^2$$
$$= \frac{0.08816}{2} \times 649^2$$
$$= 18.566\,\mathrm{kPa}$$

For gothic wing,

$$\hat{k}\frac{S}{s\,c} = \frac{4}{3} = 1.33$$

Therefore,

$$\frac{\beta}{C_f}\hat{k} = \frac{2}{1.35 \times 10^{-5}} \times 1.33$$
$$= 1.97 \times 10^5$$

For gothic wing,

$$\beta\frac{S}{c} = 0.5,\ \hat{k}\frac{L}{D} = 11.5,\ \beta C_L = 0.175$$

Therefore,

$$\frac{S}{c} = \frac{1}{4}$$
$$\mathbf{A} = \frac{4S^2}{S}$$
$$= \frac{4\,(S/c)}{\hat{k}}$$
$$= 0.75$$
$$\frac{L}{D} = 8.6$$

Concorde has four engines of thrust 4000 kg each. Actually, it is one-third of the thrust for cruise. Therefore, the drag becomes

$$D = 3 \times 4000 = 1.2 \times 10^4\,\mathrm{kgf}$$

Therefore, the thrust T is

$$T = D = 1.2 \times 10^4 \text{ kgf}$$

The power required becomes

$$\begin{aligned} D V_\infty &= (1.2 \times 10^4) \times 649 \\ &= (778.8 \times 10^4) \text{ kgf} \\ &= (778.8 \times 10^4) \times 9.81 \\ &= 7640 \times 10^4 \text{ N} \end{aligned}$$

dividing this with 746, we have power as

$$\begin{aligned} D V_\infty &= (7640 \times 10^4)/746 \\ &= 10.24 \times 10^4 \text{ hp} \\ &\approx 10^5 \text{ hp} \end{aligned}$$

The lift coefficient may be taken as $C_L \approx 0.09$. Therefore,

$$\begin{aligned} S &= \frac{L}{q_\infty C_L} \\ &= \frac{9.8 \times 10^6}{(18.566 \times 10^3) \times 0.09} \\ &\approx 5865 \text{ m}^2 \\ c^2 &= \frac{4S}{\hat{k}} \\ &= \frac{4 \times 5865}{1.33} \\ &= 17639 \text{ m}^2 \\ c &\approx 132.8 \text{ m} \end{aligned}$$

$$K_0\left(\beta \frac{S}{c}\right) = 0.7$$

The area distribution is given by

$$\frac{\bar{a}}{\overline{V}} = -\frac{3.2}{\pi}(\sin\theta - \theta\cos\theta) + \frac{3.2}{\pi}\left(\theta - \frac{\sin 2\theta}{2}\right) + \frac{2.4}{\pi}\left(\sin\theta - \frac{\sin 3\theta}{3}\right)$$

where $x/c = (1 - \cos\theta)/2$.

The elliptic vorticity load $\gamma \to \alpha_{\text{local}}(x)$, because it gives minimum drag and constant induced angle over the span.

For local angle of attack α_{local}:

$$\sigma(x) = 2\pi \left(\frac{y_1(x)}{S}\right)^2 \alpha_{\text{local}}$$
$$= \frac{1}{\pi}\frac{C_L \hat{k}}{S/c}\left[\theta - \frac{\sin 2\theta}{2} + 2\left(1 - \frac{X}{X_{cg}}\right)\left(\sin\theta - \frac{\sin 3\theta}{3}\right)\right]$$

where $\frac{y_1(x)}{S}$ is known because it is given that the wing is gothic.

1.12 Nonlinear Theories

The nonlinearity is a result of the flow separation and vortex sheet rolling, as illustrated in Fig. 1.32.

When we assume that there is no vortex sheet and its rolling, there will be singularities at the tips and the pressure coefficient would tend to become infinity ($C_p \to \infty$) there. But when there are tip vortices, the effects are nonlinear. Also,

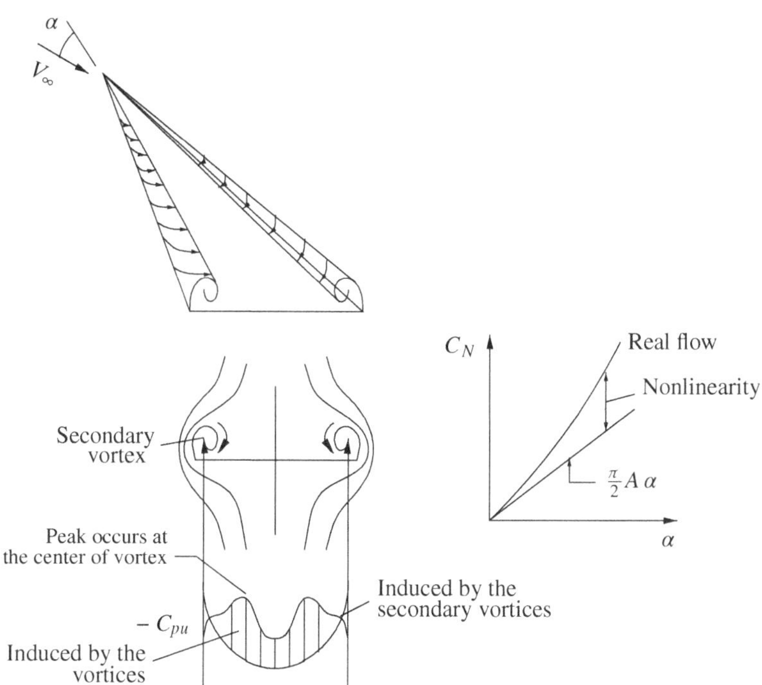

Figure 1.32 Flow separation and vortex rolling over a slender wing.

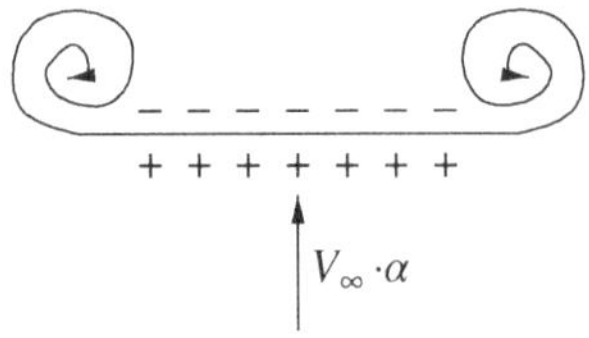

Figure 1.33 Vortex rolling over a slender wing.

the flow does not experience infinite velocity at the tips. Therefore, the pressure coefficient C_p at the tips is finite. The vortex region is a low-pressure region and so their life is more. The primary vortex induces velocity towards the tip. But the pressure is increasing toward tip (i.e. $-C_p$ decreases). Therefore, the primary flow is against the adverse pressure gradient, inducing a secondary vortex.

In addition to the primary vortex system (large ones), there will be a small secondary vortex system (in opposite direction). These two systems induce additional flow which is actually responsible for the additional lift.

There were many attempts to represent the problem by a model. Some of the popular ones are: Mangler and Smith [6], Gersten [7], and R. Hürlimann [8].

Nonlinear theory is important for off-design condition. Mangler and Smith assumed conical flow at the tips [2]. Here again one can use

$$\phi_{yy} + \phi_{zz} = 0$$

The vital requirement for solving this equation is the use of proper boundary conditions. But the problem is extremely complicated, because the boundary conditions to be applied correspond to the vortex sheet, as shown in Fig. 1.33.

There is no pressure difference across the sheet, because it is a vortex sheet, there should be no normal velocity. For this, Mangler and Smith took a number of points on the vortex sheet and computed and obtained the following result:

$$C_N = \frac{\pi}{2}\mathbf{A}\,\alpha + \left(\frac{\mathbf{A}}{4}\right)^2 4.9\left(\frac{4\,\alpha}{\mathbf{A}}\right)^{1.7} \tag{1.30}$$

This theory is valid strictly for conical flows.

1.12.1 Hürlimann Model

Hürlimann model assumed that the vortex sheet does not roll up but builds up along the chord, normal to the surface, as shown in Fig. 1.34.

There should be no functional difference across the normal vortex sheet. This model is a bit more flexible than other models in the sense that it is suitable for flows other than critical flows.

1.12.1.1 Delta Wings

For delta wings with $\alpha < \mathbf{A}/2$, the normal force coefficient is given by

$$C_N = \frac{\pi}{2}\mathbf{A}\,\alpha + \pi\,\alpha^2 \tag{1.31}$$

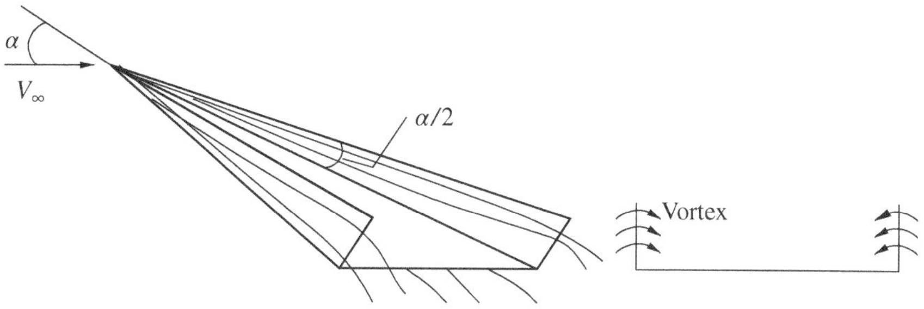

Figure 1.34 Vortex build up over a slender wing.

For delta wings with $\alpha > 5\mathbf{A}$,

$$C_N = \frac{\pi}{2}\mathbf{A}\,\alpha + 2\,\alpha^2 \tag{1.32}$$

In this result, the nonlinearity is proportional to α^2.

1.12.1.2 Gerten's Model

It can be proved that the angle made by the tip vortex with the surface is $\alpha/2$ for small aspect ratio, **A**, rectangular wings. Gerten assumed that the free vortices arising on the surface move downstream at an angle $\alpha/2$.

In the linear theory, the vortex sheet lies in a plane perpendicular to the wing. In the nonlinear theory, the vortex sheet lies at an angle of $\alpha/2$, as shown in Fig. 1.35, proved by Bollay for slender rectangular wing.

When looked from behind, all these free vortices are enclosed in a triangle whose height is $\alpha/2c$ at the trailing edge. This system can be replaced by two big vortices near the tips, because near the center of the wing, the effect is nullified due to vortices of opposite direction, as illustrated in Fig. 1.36. This can be applied to any body not necessarily to slender body alone. Gerten showed that the nonlinearity decreases with increasing aspect ratio, **A**, unlike the other theories.

Figure 1.35 Vortex sheet over a slender rectangular wing.

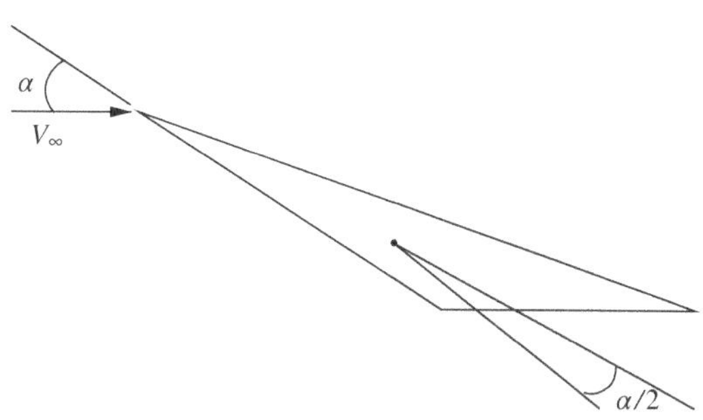

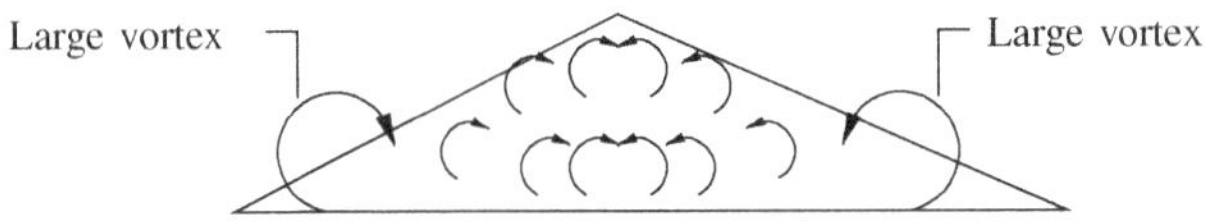

Figure 1.36 Two large opposite vortices at the trailing edge of a slender wing.

1.12.2 Breakdown of Vortices

The vortices do not extend to infinity, but breakdown after some distance.

Explanation for $C_{L_{max}}$: Variation of normal force coefficient, C_N, with the angle of attack for a slender wing is shown in Fig. 1.37.

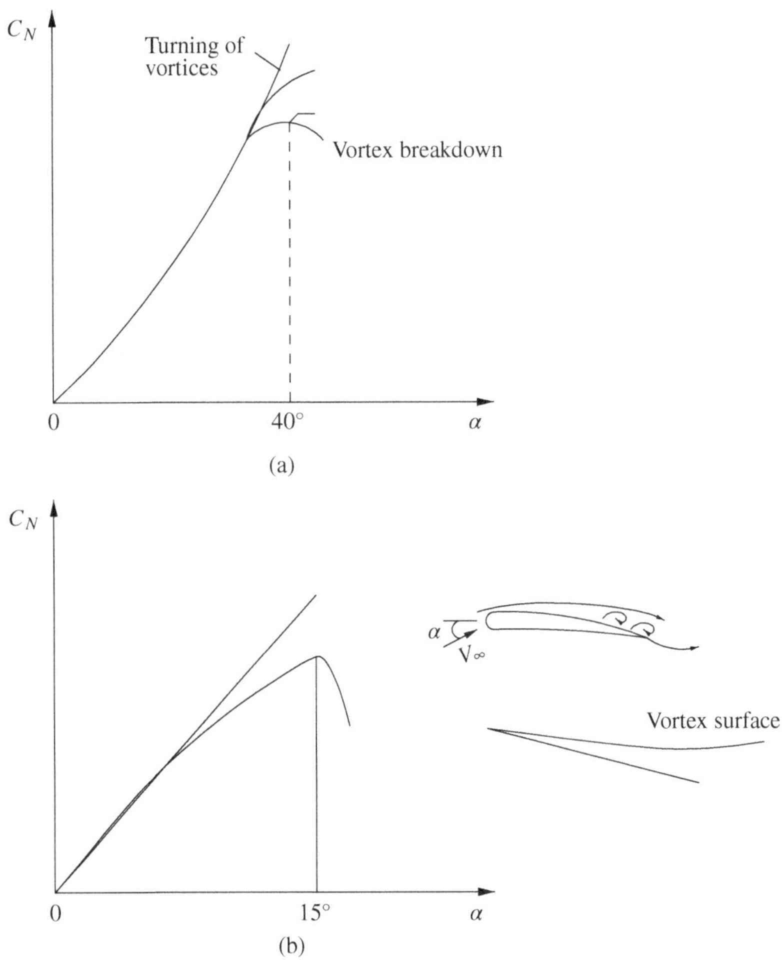

Figure 1.37 C_N variation with α. (a) for vortex breakdown and (b) for curving of vortex sheet.

The nonlinearity of C_N verses α curve is because of the following two reasons.

1. *Curving of vortex sheet*. The vortex sheet curves up after some distance as shown in Fig. 1.37(b). If this curving is over the wing itself, the nonlinearity is decreased and the induced velocity becomes lesser.
2. *Vortex breakdown*. This results in the distribution of vortices and the location of the center of the primary vortices over the wing is no more possible. This aspect further decreases the induced velocity and so the nonlinearity.

Example 1.17 Find the normal force coefficient for a missile of aspect ratio 6, flying at 5° incidence, assuming the flow around the missile as conical.

Solution

Given, $\mathbf{A} = 6$, and $\alpha = 5°$.

By Eq. (1.30), the force coefficient is

$$
\begin{aligned}
C_N &= \frac{\pi}{2}\mathbf{A}\,\alpha + \left(\frac{\mathbf{A}}{4}\right)^2 4.9\left(\frac{4\alpha}{\mathbf{A}}\right)^{1.7} \\
&= \frac{\pi}{2} \times 6 \times \frac{5}{57.3} + \left(\frac{6}{4}\right)^2 \times 4.9\left(\frac{4 \times 5}{6 \times 57.3}\right)^{1.7} \\
&= 0.822 + 11.025 \times (7.944 \times 10^{-3}) \\
&= 0.822 + 0.0876 \\
&= \boxed{0.91}
\end{aligned}
$$

1.13 Summary

Slender body configuration is a geometrical shape with one dimension much larger than the other two dimensions.

For a slender wing, the height is very large compared to the base, and the aspect ratio is much smaller than unity.

The Ogee curve is a double S-shaped curvature along the cheeks that gives the face contour and dimension (a distinctive pattern with two continuous S-shaped curves narrowing at the nose and widening at the base).

In the older type, the rockets had the control surfaces namely the fins only at the rear of it, but in the new type of rockets, the control surface runs almost up to the nose.

The popular versions of supersonic transport aircraft are the Russian TU 144, the French-British Concorde, and Boeing 2707 of the United States of America. Among these, TU 144 was designed to cruise at Mach 2.2, the Concorde at Mach 2, and the Boeing 2707 at Mach 2.7.

Currently Boom Technology, Inc. is developing supersonic transport aircraft. Boom Technology is designing Mach 1.7, 65-88-passenger supersonic airliner. It is claimed that the maximum noise level of this aircraft is only around 60 dB.

The wings meant for supersonic flight should have the flow characteristics such that they are suitable for supersonic aircraft.

The drag caused by the waves is termed as *wave drag* which is a component of pressure drag due to compressibility effects. It is caused by the formation of shock waves around a body.

Skin friction drag is the drag caused by the fluid shearing along the surface of a body.

The center of pressure (the point at which the line of action of the resultant of the forces due to pressure acts) is at the quarter chord point $c/4$ for the plate in subsonic stream and at the midpoint $c/2$ in the supersonic stream.

A slender wings' behavior is independent of freestream Mach number, therefore, there is no trim problem. It is the component of the freestream Mach number M_∞ that is normal to the leading edge determines the behavior of the wing.

By definition, slender wings have subsonic leading edges.

In off design conditions, there will be leading edge separation, that is, when the angle of attack α is high, the flow has to go over the leading edge very quickly and it can no more go along the surface. The flow will get separated, curl, and form small vortices behind the leading edge.

The vortex breakdown experienced is a stability problem. At some point, the vortex breaks down and this is a troublesome phenomenon because the breakdown point is far ahead of the trailing edge.

At the design condition, there will be no vortices when the flow hits the leading edge tangentially (this is achieved by cambering the wing).

Providing sharp leading edge to a slender wing with the aim of making the wing not to experience separation at the leading edge.

The ground effect (the effect of aerodynamic buoyancy produced by a cushion of air below a vehicle moving close to the ground) encountered by a slender wing is strongly influenced by the aspect ratio.

Viscous effects become very important because it makes the boundary layer to become so thick.

The basic equation for all classical linearized theories is

$$(1 - M_\infty^2)\,\phi_{xx} + \phi_{yy} + \phi_{zz} = 0$$

For slender bodies, the changes in velocity along the length of the body is much smaller than the velocities in the other directions.

The pressure coefficient at any point on the body can be expressed as

$$C_p = \frac{p - p_\infty}{\frac{1}{2}\rho_\infty V_\infty^2}$$

The transverse flow field will be proportional to $V_\infty \alpha$, where V_∞ is the freestream velocity, and α is the angle of attack. The governing equation is

$$\boxed{\Delta\phi_\alpha = 0}$$

This represents lift problem.

The governing equation for longitudinal flow at zero angle of attack, $\alpha = 0$, is

$$\boxed{\Delta\phi_0 = 0}$$

Most of the early developments in supersonic missile aerodynamics started with the assumption of inviscid flow that was given a small perturbation by the movement of the missile. If the body were sufficiently long and thin, the "slender body" theory could be used to describe the perturbed flow. A linearized partial differential equation, the potential equation, described the flow field. The solution for this simplified model yields the estimates of dimensionless coefficients of lift (C_L) and drag due to lift (C_{D_L}) that are independent of the flight Mach number and body shape (within the slenderness restrictions of the theory) and of the static pitching moment (C_M) that depends on missile volume.

The general solution for flow past a slender body can be expressed as

$$\phi(x, y, z) = \phi_0(x, y, z) + \phi_\alpha(x, y, z) + g(x,\ M_\infty)$$

For a wing without camber and without thickness, the sectional lift coefficient is

$$dL = 2\pi q_\infty\, d(\alpha y_1^2)\, dx$$

where $\alpha = \alpha(x)$ and $q_\infty = \frac{1}{2}\rho_\infty V_\infty^2$.

Kutta condition states that the flows from the upper and the lower portions of an airfoil rejoin at the trailing edge with no sudden change in the flow direction. Kutta condition is fully satisfied for ogival and gothic wings that are more commonly used than the delta shaped wings. For this case,

$$C_L = \frac{\pi}{2}\mathbf{A}\,\alpha$$

The pitching moment, M, for the slender body is

$$M = -\int_0^c (x - x_{cg})\, dL$$

Pitching moment coefficient becomes

$$\begin{aligned} C_M &= \frac{M}{q_\infty S C_\mu} \\ &= \frac{-2\,\pi\,\alpha}{S C_\mu}\left[(c - x_{cg})S^2 - \int_0^c y_1^2 dx\right] \end{aligned}$$

where

$$C_\mu = \frac{1}{S}\int_{-s}^{+s} c^2 dy$$

This C_M cannot be expressed by a general relation.

The dynamic derivatives are necessary to study the dynamic stability.

For steady flow, the pressure coefficient is

$$C_p = -\frac{2}{V_\infty}\left(\frac{\partial\phi}{\partial x}\right)$$

The pressure coefficient for unsteady flow is

$$C_p = -\frac{2}{V_\infty}\left(\frac{\partial\phi}{\partial x} + \frac{1}{V_\infty}\frac{\partial\phi}{\partial t}\right)$$

For Delta wing:

$$C_L = \left(\frac{\pi}{4}\mathbf{A}\right)\left(\frac{\dot{\alpha}\, C_\mu}{V_\infty}\right)$$

$$C_{L\dot{\alpha}} = \frac{\partial C_L}{\partial\left(\frac{\dot{\alpha}\, C_\mu}{V_\infty}\right)} = \frac{\pi}{4}\mathbf{A}$$

In general, the pressure coefficient may be expressed as

$$C_p = \pm\frac{2}{V_\infty}\dot{\alpha}\sqrt{y_1^2 - y^2} \neq 0\,, \quad \text{when}\, y = 0$$

Longitudinal flow problem is a problem of source–sink distribution. For this problem, in general, the velocity inside will not be zero. Therefore, to make the flow velocity zero a vortex distribution has to be superposed, so that the resultant velocity inside becomes zero.

For bodies with sharp corners, there is no problem in solving the flow field in principle, because only the potential function ϕ has to be chosen suitably and rest all the same as before.

For a slender body in an inviscid flow, to have a finite C_N (i.e. to have $C_N \neq 0$), there should be a base span. For such bodies, C_N will be the same for any planform. The moment coefficient is given by

$$C_M = 4\frac{1}{Sc_\mu}\pi\,\alpha\int_0^c (\alpha - x_{cg})\, d\left(\frac{y_1^2}{2}\right)$$

Drag due to friction is

$$\frac{D_f}{q_\infty} \approx 2\overline{C}_f\, S$$

The total drag is

$$\frac{D}{q_\infty} = \frac{T}{q_\infty} + \frac{D_f}{q_\infty}$$

$$= \frac{128}{\pi} k_0 \frac{\mathbf{V}^2}{c^4} + 2\,\overline{C}_f \hat{k} s\, c$$

where $\hat{k} = S/(c\,s)$. This depends on the planform shape. For optimum drag,

$$\frac{\partial (D/q_\infty)}{\partial c} = 0$$

which gives

$$\boxed{C_{D_{\text{opt}}} = \left[\frac{128}{\pi} k_0 \frac{\mathbf{V}^2}{2\,\overline{C}_f \hat{k} \frac{S}{c}}\right]^{1/6}}$$

k_0 depends on $A(x)$.

The induced drag is a minimum if $\gamma(\eta)$ is an elliptic distribution (i.e. elliptic distribution for the span-wise distribution of lift).

The nonlinearity is a result of the flow separation and vortex sheet rolling. When we assume that there is no vortex sheet and its rolling, there will be singularities at the tips and the pressure coefficient would tend to become infinity ($C_p \to \infty$) there. But when there are tip vortices, the effects are nonlinear. Also, the flow does not experience infinite velocity at the tips. Therefore, the pressure coefficient C_p at the tips is finite. The vortex region is a low-pressure region and so their life is more. The primary vortex induces velocity toward the tip. But the pressure is increasing towards tip (i.e. $-C_p$ decreases). Therefore, the primary flow is against the adverse pressure gradient, inducing a secondary vortex.

For delta wings with $\alpha < \mathbf{A}/2$, the normal force coefficient is given by

$$C_N = \frac{\pi}{2}\mathbf{A}\,\alpha + \pi\,\alpha^2$$

For delta wings with $\alpha > 5\mathbf{A}$,

$$C_N = \frac{\pi}{2}\mathbf{A}\,\alpha + 2\,\alpha^2$$

In this result, the nonlinearity is proportional to α^2.

Exercise Problems

1.1 The aerodynamic efficiency of a slender body is 2.6. If the drag coefficient due to skin friction is 0.4 and the drag coefficient caused by viscous friction is 0.1, find the lift coefficient associated with this mode of flight.

[Ans. 1.3]

1.2 Determine the limiting value of the nose angle for a conical slender body flying at Mach 2 not to experience wave drag.

[Ans. 60°]

1.3 Determine the limiting maximum Mach number up to which a slender two-dimensional wedge of nose angle 10° can fly without encountering wave drag.

[Ans. 11.47]

1.4 The pitching moment coefficient for a slender body shown in Fig. 1.12, at an angle of incidence 3°, is –0.2. If the average chord is 2 m and the center of gravity at 0.5 chord, determine the aspect ratio of the body.

[Ans. 9.73]

1.5 A flat plate of length 3 m and width 2 m flies at Mach 2 with 5° incidence at the sea level. If the rotation about the center of gravity is 50 radian/s, determine the *cg* location from the leading edge.

[Ans. 1.814 m]

1.6 If a slender wing of span 4 m and chord 1 m flies at Mach 1.5 at 2° angle of attack at sea level, determine the lift generated.

[Ans. 139.6 kN]

1.7 A symmetrical semi-wedge of nose angle 10° and length 3 m flies at sea level. If the freestream Mach number is 2.2, find the normal component of the perturbation velocity at the peak point. located at 1.5 m from the nose.

[Ans. 129.70 m/s]

1.8 If a missile of volume 0.4 m^3, base radius 0.3 m, length 4 m experiences a pitching moment of –0.12, determine the angle of incidence.

[Ans. 5.32°]

1.9 A slender body of triangular shape with base 1 m, length 6 m experiences a pitching moment of –0.25. If the center of gravity is at 30% of the length from the vertex, determine the angle of incidence.

[Ans. 2.76°]

1.10 If the aerodynamic efficiency of a cylindrical slender body of diameter 0.8 m, total length 4 m, with nose-cone of length 1 m is 25, determine the pitching moment coefficient about the nose tip.

[Ans. –0.507]

2

Hypersonic Aerodynamics (Slender Bodies)

2.1 Introduction

In gas dynamic theory of perfect gas, hypersonic flow is defined as the flow with Mach number greater than 5, where the change in flow Mach number, M, is dictated by the change in the speed of sound, a. That is, in hypersonic flow regime, speed of sound, a, is more dominant than the flow speed, V, itself. But in problems such as flow around blunt bodies begin to exhibit many of the characteristics of hypersonic flow, when the Mach number is 4 or greater. By definition,

$$\boxed{M_\infty \equiv \frac{V_\infty}{a_\infty} \gg 1} \tag{2.1}$$

where M_∞ is the freestream Mach number, V_∞ is the freestream flow speed, and a_∞ is the freestream speed of sound. That is, the Mach number is greatly larger than unity ($M_\infty \gg 1$) is the basic assumption for all hypersonic flow theories. Thus, the internal thermodynamic energy of the freestream fluid particles is small compared to the kinetic energy of the freestream, for hypersonic flows. In flight applications, this results because the volume of the fluid particles is relatively large. The limiting case, where M_∞ approaches infinity because the freestream velocity approaches infinity while the freestream thermodynamic state remains invariant, produces extremely high temperatures in the shock layer.

The high temperatures associated with hypersonic flight are difficult to match in ground-test facilities, such as hypersonic wind tunnel, and shock tunnel. Therefore, in wind tunnel applications, hypersonic Mach numbers are achieved through relatively low speeds of sound. Thus, in hypersonic wind tunnels, the test-section

Hypersonic Slender Body Aerodynamics, First Edition. Ethirajan Rathakrishnan.

Companion Website: www.wiley.com/go/rathakrishnan/hypersonic

Mach number, M_∞, approaches infinity because the speed of sound goes to *zero* while the freestream velocity is held constant. As a result, the fluid temperature in such wind tunnels remains below the level that would damage the wind tunnel model.

Another assumption common to hypersonic flow is that the ratio of the freestream density to the density just behind a shock is extremely small, that is,

$$\boxed{\epsilon \equiv \frac{\rho_\infty}{\rho_2} \ll 1} \tag{2.2}$$

where ρ_∞ is the freestream density and ρ_2 is the density behind the shock. Equation (2.2) is known as the *small-density ratio* assumption. Thus, this assumption primarily relates the properties of the gas downstream of the shock wave (specified with subscript 2) to freestream (specified by subscript ∞) properties. The ratio of density, ρ_1, ahead of and behind, ρ_2, of a normal shock in a freestream flow with $M_\infty \rightarrow \infty$ can be expressed as [1]

$$\epsilon \equiv \frac{\rho_1}{\rho_2} = \frac{\gamma - 1}{\gamma + 1} \tag{2.3}$$

Thus, ϵ is 1/6 for perfect gases with specific heats ratio $\gamma = 1.4$. It is important to note that a typical hypersonic wind tunnel that operates at conditions where the test gas can be treated as perfect gas. However, during re-entry flights, the density ratio, ρ_∞/ρ_2, can approach value as low as 1/20. Thus, the density ratio simulation in those wind tunnels where the airflow behaves as a perfect gas does not match the flight values. This will have a significant effect on the shock standoff distance, as well as on other flow parameters. To generate low-density ratios for wind tunnel simulations, gases other than air, such as hexafluoroethane (C_2F_6) for which $\gamma = 1.1$, or tetrafluoromethane (CF_4), also known as carbon tetrafluoride, for which $\gamma = 1.2$, may be used as the test gas.

Example 2.1 A missile is launched vertically upward at the sea level. If the speed at launch is 1500 m/s, determine the altitude at which the flight Mach number will become 5.

Solution

Given, $V = 1500$ m/s. At the sea level, $T = 15°\text{C} = 273.15 + 15 = 288.15$ K.

Therefore, the speed of sound is

$$\begin{aligned} a &= \sqrt{\gamma R T} \\ &= \sqrt{1.4 \times 287 \times 288.15} \\ &= 340.26\ \text{m/s} \end{aligned}$$

The Mach number is

$$M = V/a$$
$$= 1500/340.26$$
$$= 4.4$$

For Mach 5, the speed of sound is

$$5 = V/a$$
$$a = V/5$$
$$= 1500/5$$
$$= 300\,\text{m/s}$$

Therefore, the temperature T becomes

$$300 = \sqrt{\gamma R T}$$
$$300^2 = (1.4 \times 287)\,T$$
$$T = 300^2/(1.4 \times 287)$$
$$= 224\,\text{K}$$

From atmospheric table, for $T = 224$ K, the altitude is $h = \boxed{9800\,\text{m}}$.

Example 2.2 A model has to be tested in a hypersonic tunnel, at a freestream temperature of 200 K, with helium as the test gas at Mach 9. Treating helium as a perfect gas, determine the total temperature of the gas required in the settling chamber.

Solution

Given, $M = 9$, and $T = 200$ K. For helium, the specific heats ratio γ is given by

$$\gamma = \frac{n+2}{n}$$

where n is the number of degree of freedom. For helium, being a monatomic gas, $n = 3$, therefore,

$$\gamma = \frac{3+2}{3}$$
$$= 1.67$$

For $M = 9$, from gas table,

$$\frac{T}{T_0} = 0.035543$$

Therefore, the total temperature, T_0, in the settling chamber becomes

$$T_0 = \frac{T}{0.035543} = \frac{200}{0.035543} = \boxed{5627\ \mathrm{K}}$$

2.2 Mach Number Independence Principle

For slender body configurations, such as sharp-nosed cones and wedges, the strong shock assumption is

$$\boxed{M_\infty \sin \theta_b \gg 1} \tag{2.4}$$

where M_∞ is the freestream Mach number, and θ_b is the semi-angle of the nose. The concept termed as the *Mach number independence principle*[1] depends on this assumption.

The Mach number independence principle was derived, for inviscid flow, by Oswatitsch [7]. For blunt bodies or slender bodies, at large angles of incidence, the pressure forces are much larger than the viscous forces, when the Reynolds number exceeds 10^5. Therefore, one would expect the Mach number independence principle to hold at these conditions.

Koppenwallner [9] demonstrates that there is a significant increase in the total drag coefficient for right circular cylinder due to the friction drag when the Knudsen number, Kn, defined as the ratio of mean-free path length, λ, to a characteristic dimension, d, in the flow field, is greater than 0.01. Using the Reynolds number, based on the flow conditions behind a normal shock wave; ρ_2, V_2, and μ_2 are given by

$$\mathrm{Re}_2 = \frac{\rho_2 V_2 d}{\mu_2} \tag{2.5}$$

as the characteristic parameter, the skin friction drag coefficient, in terms of Re_2, is given by

$$\boxed{C_{Df} = \frac{5.3}{(\mathrm{Re}_2)^{1.18}}} \tag{2.6}$$

1 The hypersonic Mach number independence principle of Oswatitsch is important for hypersonic vehicle design. It explains why, above a certain flight Mach number (M_∞ = 4-6, depending on the body shape), some aerodynamic properties become independent of the flight Mach number. For ground test facilities, this means that it is sufficient for the Mach number in the test section to be high enough that Mach number independence exists.

Example 2.3 Find the skin friction drag coefficient of the slender body of characteristic length 4 m flies at Mach 3 at 3000 m altitude.

Solution

Given, $l = 4$ m, $M = 3$, and $h = 3000$ m.

At 3000 m altitude, from atmospheric table, $T = 268.65$ K, $\rho = 0.90929$ kg/m^3.

There will be a detached shock at the nose of the body. Treating the portion of the wave ahead of the nose as normal shock, we have, for $M_1 = 3$, from normal shock table, $M_2 = 0.4752$, $\rho_2/\rho_1 = 3.8571$, and $T_2/T_1 = 2.679$. Therefore, the density, ρ_2, and temperature, T_2, at the nose, are

$$\begin{aligned}\rho_2 &= 3.8571\,\rho_1\\ &= 3.8571 \times 0.90929\\ &= 3.507\ \text{kg/m}^3\\ T_2 &= 2.679\,T_1\\ &= 2.679 \times 268.65\\ &= 719.71\ \text{K}\end{aligned}$$

The speed of sound, a_2, flow speed, V_2, and the viscosity coefficient, μ_2, behind the shock are,

$$\begin{aligned}a_2 &= \sqrt{\gamma R T_2}\\ &= \sqrt{1.4 \times 287 \times 719.71}\\ &= 537.75\ \text{m/s}\\ V_2 &= M_2\,a_2\\ &= 0.4752 \times 537.75\\ &= 255.54\ \text{m/s}\\ \mu_2 &= (1.46 \times 10^{-6}) \times \frac{T^{3/2}}{T + 111}\\ &= (1.46 \times 10^{-6}) \times \frac{719.71^{3/2}}{719.71 + 111}\\ &= (1.46 \times 10^{-6}) \times \frac{19307.96}{830.71}\\ &= 3.39 \times 10^{-5}\ \text{kg/(m s)}\end{aligned}$$

The Reynolds number of the flow behind the shock at the nose is

$$\begin{aligned}\text{Re}_2 &= \frac{\rho_2 V_2 c}{\mu_2}\\ &= \frac{3.507 \times 255.54 \times 4}{3.39 \times 10^{-5}}\\ &= 1057.44 \times 10^5\end{aligned}$$

The skin friction drag coefficient, by Eq. (2.6), is

$$\begin{aligned}C_{Df} &= \frac{5.3}{(\mathrm{Re}_2)^{1.18}} \\ &= \frac{5.3}{(1057.44 \times 10^5)^{1.18}} \\ &= \frac{5.3}{2.94 \times 10^9} \\ &= \boxed{1.8 \times 10^{-9}}\end{aligned}$$

2.3 Atmospheric Properties

For calculating the aerodynamic forces acting on a flying vehicle, it is necessary to know the atmospheric conditions prevailing around it. Since the atmospheric condition is not constant with altitude, a model must be developed to account for the changes in atmospheric condition with altitude. Wind also plays an important role in the flight of a rocket, and therefore it is important to use a realistic wind model in the simulation.

2.3.1 Atmospheric Model

The atmospheric properties that are of interest are the density of air, ρ, which is a scaling parameter to the aerodynamic coefficients via the dynamic pressure, $q = \frac{1}{2}\rho V^2$, and the speed of sound, a, which affects the Mach number, M, of the rocket, which in turn affects its aerodynamic properties. These may in turn be calculated from the air pressure, p, and temperature, T.

Several models exist that define standard atmospheric conditions as a function of altitude, including the International Standard Atmosphere (ISA) [10] and the US Standard Atmosphere [11]. These models yield identical temperature and pressure profiles for altitudes up to 32 km.

These models are based on the assumption that air follows the ideal gas law,

$$\rho = \frac{mp}{RT}$$

where m is the molecular mass of air, and R is the ideal gas constant, 8.314 J/(mol K). From the equilibrium of hydrostatic forces, the differential equation for pressure as a function of altitude z can be found as

$$\begin{aligned}dp &= -g_0\,\rho\,dz \\ &= -g_0\,\frac{mp}{RT}\,dz\end{aligned}$$

where g_0 is the gravitational acceleration. If the temperature of air were to be assumed to be constant, this would yield an exponential diminishing of air pressure.

The ISA and US Standard Atmospheres further specify a standard temperature and pressure at the sea level and a temperature profile for the atmosphere. The temperature profile is given as eight temperatures for different altitudes, which are then linearly interpolated. The temperature profile and base pressures for the ISA model are presented in Table A.2, in the Appendix.

These models are totally static and do not take into account any local fight condition.

Example 2.4 Determine the molecular mass of air at the standard sea level state.

Solution
At the sea level, the density, temperature, and pressure of air are $\rho = 1.225\ \text{kg/m}^3$, $T = 15°\text{C} = 15 + 273.15 = 288.15\ \text{K}$, and $p = 101,325\ \text{Pa}$.

By ideal gas law,

$$\rho = \frac{mp}{RT}$$

where $R = 8.314\ \text{J/(mol K)}$. Therefore, the molecular mass becomes

$$\begin{aligned} m &= \frac{\rho RT}{p} \\ &= \frac{1.225 \times 8.314 \times 288.15}{101\,325} \\ &= \boxed{0.02896\ \text{kg/mol}} \end{aligned}$$

2.4 Hypersonic Flow Characteristics

The characteristics of hypersonic flow can be computed using the fluid dynamic phenomena, when the configuration geometry, the attitude or orientation of the geometry to the freestream flow, and the altitude at which the vehicle flies or the test conditions of a wind tunnel simulation are known. When computer-generated flow field solutions are used to define the aerothermodynamic environment, matching the fluid-dynamic similarity parameters becomes a secondary issue. However, it is essential to incorporate *critical* fluid-dynamic phenomena into the computational flow model. Now it is essential to note that a typically important fluid-dynamic phenomenon for one application may be irrelevant to another. For example, the drag coefficient for a cylinder whose axis is perpendicular to a hypersonic freestream will be essentially constant, independent of Mach number and Reynolds number, provided that the Mach number is sufficiently large so that the flow is hypersonic and the Reynolds number is sufficiently large so that the boundary layer is thin. For blunt bodies, where skin friction is a small fraction of the total drag, reasonable estimates of the force coefficient could be obtained from flow fields computed using flow models based on the Euler equation, that is, neglecting viscous terms in the equations of motion. However, in the rarefied (low density) flows encountered at higher altitudes, viscous/inviscid interactions become important and the effects of viscosity can no longer be neglected.

2.5 Governing Equations

For hypersonic flow ($M_\infty > 5$) past slender bodies, the changes in flow properties are continuous; however, there are some perturbations. Therefore, it is not possible to linearize the equations as in supersonic flow. This is because the fundamental equation governing most of the compressible flow regime [12]

$$(1 - M_\infty^2)\,\phi_{xx} + \phi_{yy} + \phi_{zz} = 0 \tag{2.7}$$

valid for subsonic, transonic, and supersonic flow under the framework of small perturbations with $u \ll V_\infty$, $v \ll V_\infty$, and $w \ll V_\infty$ is not valid for hypersonic flow even for slender bodies, since $u \approx V_\infty$ in the hypersonic flow regime.

Thus, the linear equation governing supersonic flow becomes nonlinear for hypersonic flows, because for hypersonic flow past slender bodies, the assumption that the components of perturbation velocity, u, v, w, are small compared to freestream flow velocity, V_∞, and freestream speed of sound a_∞ is not valid. This is because for $M_\infty > 5$: (i) $V_\infty \approx 5a_\infty$ and (ii) $u \approx a_\infty/10$. Therefore, the second condition is satisfied even though the first condition is not violated.

It is well known that for nonlinear flows, we have to deal with the shocks unlike the linearized flows where the shocks can be replaced with Mach waves. In the case of hypersonic flows, even for the case of slender bodies, the shocks present cannot be assumed to be Mach waves, as in the case of supersonic flows.

The shocks over the body in a hypersonic flow are curved and very close to the body, as shown in Fig. 2.1. Shocks close to the body surface has very serious implications, because the boundary layer, irrespective of its thickness will affect the flow considerably. This viscous interaction makes it impossible to treat the problem as consisting of separate inviscid (flow outside the boundary layer) and boundary layer effect.

To solve the flow field between the bow shock wave and the surface of a vehicle flying at hypersonic speed, it is necessary to develop the governing equations of motion and the appropriate flow models.

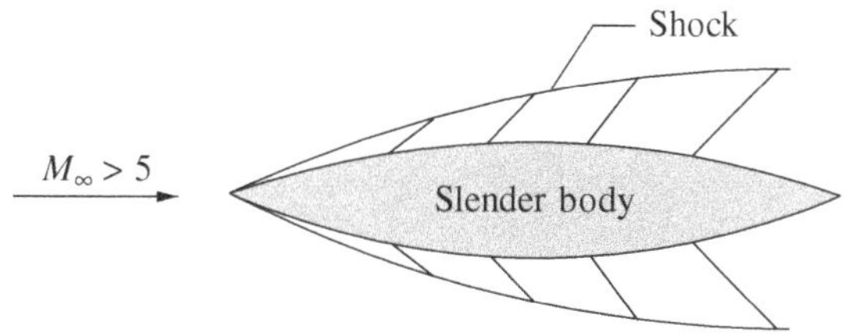

Figure 2.1 Flow past a slender body in hypersonic flow.

Let us consider the flow past a body moving at a hypersonic speed as shown in Fig. 2.1. Note that the entropy change across a shock wave depends on the freestream Mach number and the shock inclination angle. Thus, the entropy change across the bow shock wave depends on where the flow crosses the shock wave. The flow in the inviscid portion of the shock layer, that is, the flow between the shock wave and the boundary layer, is rotational. The flow downstream of that portion of the shock wave where the shock angle is relatively low, rendering the shock as weak, the flow is supersonic, in accordance with shock theory that the flow past a weak shock will be supersonic, with Mach number less than the upstream value.

Examine the flow field over the body. The flow process could be a chemical reaction or the exchange of energy among the various modes, for example, translational, rotational, vibrational, and electronic energies, of the atom and molecule. The transfer of energy between the various energy modes is accomplished through the collisions between the molecules and atoms. At high altitudes, where the air density is low, chemical states need not necessarily reach equilibrium. Nonequilibrium processes occur in a flow when the time required for a process to accommodate itself to the local conditions within a particular region is of the same order as the time it takes for the air particles to cross that region. If the time required for accommodation is longer than the transit time, the chemical composition remains "frozen" as the flow moves around the vehicle.

One usually goes in for interaction method to solve this problem. This involves procedure for solving the problem considering the flow inside the boundary layer and the influence of boundary layer on the flow outside of it. There will be dissociation and therefore the gas ceases to be perfect.

In reality, for hypersonic flight, the temperature of the gas molecules that pass through the detached shock wave increases to very high levels, leading to the excitation of vibrational and chemical energy modes. This excitation lowers the specific heats ratio of the gas below the freestream value, if it is assumed that equilibrium exists and that dissociation is not driven to completion. A large amount of energy that would have gone into increasing the static temperature behind the bow shock wave for a perfect gas is used instead to excite the vibrational energy levels or to dissociate the gas molecules, in hypersonic flows which renders the flow as a high-enthalpy stream that ceases to be perfect. At the sea level, at about 900 m/s, the vibrational energy of the air molecules begins to become important. Oxygen dissociation begins when the freestream velocity is in the range from 1800 m/s to 2400 m/s. Nitrogen dissociation occurs when the freestream velocity exceeds 4500 m/s. When the freestream speed exceeds 9000 m/s, atoms get ionized. Note that the dissociation and ionization reactions are pressure dependent because each molecule yields two product particles, and such reactions are inhibited by high pressure. Therefore, high temperature and, consequently, higher velocity are required to produce the reaction at the sea level than at high altitude where the pressure is much lower than the sea level pressure.

To account for the departure from the thermally perfect equation of state, $p = \rho RT$, due to the chemical reaction in air, the compressibility factor z is introduced. The compressibility factor is the ratio of the molecular weight of the undissociated air to the mean molecular weight at the conditions of interest [12]. Thus,

$$z = \frac{M_o}{\overline{M}} \tag{2.8}$$

where $\overline{M}$ is the mean molecular weight of the gas mixture at the conditions of interest. Accounting for the change in the gas composition, the state equation becomes

$$\begin{aligned} p &= \rho \frac{R_u}{\overline{M}} T \\ &= \rho \frac{M_o}{\overline{M}} \frac{R_u}{M_o} T \end{aligned}$$

But $M_o/\overline{M} = z$ and $R_u/M_o = R$, therefore,

$$p = \rho z RT \tag{2.9}$$

At this stage, it is essential to note that the compressibility factor, z, is not influenced by the vibrational excitation and, therefore, $z = 1$ until the dissociation of oxygen begins. Since air contains about 20% oxygen, $z \to 1.2$ as the oxygen dissociation approaches completion. Similarly, $z \to 2.0$ as the nitrogen dissociation approaches completion and all the molecules have dissociated into atoms. The ionization process produces further increase in z.

As energy is absorbed by the gas molecules entering the shock layer, the conservation laws and their thermophysics induce certain changes in the forebody flow (i.e. the flow just ahead of the nose). The static temperature, the speed of sound, and the velocity in the shock layer are lesser for the equilibrium, real gas flow than for a perfect gas flow. The static pressure for air in the thermodynamic equilibrium is slightly larger than the perfect gas value. The density is increased considerably, and, as a result, the shock layer thickness is reduced significantly.

Computation of properties changes across a normal shock wave, using thermochemical equilibrium air properties [12] shows that, for temperatures of 1600 K and below, the air composition does not change in the shock-compression process. Therefore, for temperatures below 1600 K, the gas can be treated as thermally perfect. Furthermore, for temperatures below 800 K, it can be seen that the air can be treated as perfect with constant specific heats ratio of 1.4.

Example 2.5 Determine the ratio of perturbation velocity to freestream velocity in the streamwise direction of the slender body vehicle that cruises at Mach 8 at 5 km altitude.

Solution

Given, $M = 8$, and $h = 5$ km. At 5 km altitude, from standard atmospheric table, the temperature is $T = 255.65$ K. Therefore, the speed of sound in the freestream is

$$\begin{aligned} a &= \sqrt{\gamma R T} \\ &= \sqrt{1.4 \times 287 \times 255.65} \\ &= 320.5 \text{ m/s} \end{aligned}$$

For $M = 8$, the freestream velocity is $V \approx 8\,a$ and the perturbation velocity is $u \approx a/10$. Therefore,

$$\begin{aligned} V &\approx 8 \times 320.5 \\ &\approx 2564 \text{ m/s} \\ u &\approx 320.5/10 \\ &\approx 32.05 \text{ m/s} \end{aligned}$$

Thus,

$$\begin{aligned} \frac{u}{V} &= \frac{32.05}{2564} \\ &= \boxed{0.0125} \end{aligned}$$

Example 2.6 If the compressibility factor of air is 0.82, determine the molecular weight of the undissociated air.

Solution

Given, $z = 0.82$. The compressibility factor z is given by

$$z = \frac{M_0}{M}$$

For dry air, the mean molecular weight $\overline{M}$ is 28.96 unit. Therefore, the weight of the undissociated air M_o is

$$
\begin{aligned}
M_o &= z \times \overline{M} \\
&= 0.82 \times 28.96 \\
&= \boxed{23.747\,\text{unit}}
\end{aligned}
$$

2.6 Comparison of Supersonic and Hypersonic Flow

To gain an understanding about the dominance of change in sound speed, a, over the flow speed, V, let us have a closer look at the changes in a and V associated with accelerating a supersonic flow from Mach 2 to 3 and a hypersonic stream from Mach 6 to 7. Let the gas be air and the static temperature of Mach 2 and 6 streams is the same at 218 K.

Supersonic flow:

Given, $M_1 = 2$, $T_1 = 218$ K, and $M_2 = 3$. Let us assume that both the accelerations are isentropic, and the specific heats ratio $\gamma = 1.4$.

For the given initial state of Mach 2 flow at 218 K, the total temperature, T_{01}, given by isentropic relation, is

$$
\begin{aligned}
T_{01} &= T\left(1 + \frac{\gamma - 1}{2} M_1^2\right) \\
&= 218 \times \left(1 + \frac{1.4 - 1}{2} \times 2^2\right) \\
&= 392.4\,\text{K}
\end{aligned}
$$

The speed of sound is

$$
\begin{aligned}
a_1 &= \sqrt{\gamma R T_1} \\
&= \sqrt{1.4 \times 287 \times 218} \\
&= 295.96\,\text{m/s}
\end{aligned}
$$

The flow speed is

$$
\begin{aligned}
V_1 &= M_1 a_1 \\
&= 2 \times 295.96 \\
&= 591.92\,\text{m/s}
\end{aligned}
$$

For the accelerated stream, $M_2 = 3$, $T_{02} = T_{01} = 392.4$ K. Therefore, the static temperature becomes

$$
T_2 = \frac{T_{02}}{1 + \frac{\gamma - 1}{2} M_2^2}
$$

$$= \frac{392.4}{1 + \frac{1.4-1}{2} \times 3^2}$$
$$= 140.14\,\text{K}$$

The speed of sound is

$$a_2 = \sqrt{\gamma R T_2}$$
$$= \sqrt{1.4 \times 287 \times 140.14}$$
$$= 237.29\,\text{m/s}$$

The flow speed is

$$V_2 = M_2 a_2$$
$$= 3 \times 237.29$$
$$= 711.87\,\text{m/s}$$

The change in flow speed due to acceleration from Mach 2 to 3 is

$$\Delta V = V_2 - V_1$$
$$= 711.87 - 591.92$$
$$= 119.95$$

The change in speed of sound associated with M change from 2 to 3 is

$$\Delta a = a_2 - a_1$$
$$= 237.29 - 295.96$$
$$= -58.67$$

The percentage (%) change in V and a associated with M change from 2 to 3 are

$$\frac{\Delta V}{V_1} \times 100 = \frac{119.95}{591.92} \times 100$$
$$= \boxed{20.26\%}$$
$$\frac{\Delta a}{a_1} \times 100 = \frac{-58.67}{295.96} \times 100$$
$$= \boxed{-19.82\%}$$

Hypersonic flow:

Given, $M_1 = 6$, $T_1 = 218$ K, and $M_2 = 7$.

Assuming that the acceleration from Mach 6 to 7 is isentropic and $\gamma = 1.4$, we have the stagnation temperature as

$$T_{01} = T_1 \left(1 + \frac{\gamma - 1}{2} M_1^2\right)$$

$$= 218\left(1 + \frac{1.4 - 1}{2} \times 6^2\right)$$
$$= 1787.6\,\text{K}$$

The speed of sound, a_1, and flow speed, V_1, are

$$a_1 = \sqrt{\gamma R T_1}$$
$$= \sqrt{1.4 \times 287 \times 218}$$
$$= 295.96\,\text{m/s}$$
$$V_1 = M_1 a_1$$
$$= 6 \times 295.96$$
$$= 1775.76\,\text{m/s}$$

At the accelerated state of $M_2 = 7$, the static temperature is

$$T_2 = \frac{T_{02}}{1 + \frac{\gamma - 1}{2} M_2^2}$$
$$= \frac{1787.6}{1 + \frac{1.4 - 1}{2} \times 7^2}$$
$$= 165.52\,\text{K}$$

The speed of sound, a_2, and flow speed, V_2, are

$$a_2 = \sqrt{\gamma R T_2}$$
$$= \sqrt{1.4 \times 287 \times 165.52}$$
$$= 257.89\,\text{m/s}$$
$$V_2 = M_2 a_2$$
$$= 7 \times 257.89$$
$$= 1805.23\,\text{m/s}$$

The percentage (%) change in V and a associated with Mach number, M, change from 6 to 7 are

$$\frac{\Delta V}{V_1} \times 100 = \frac{V_2 - V_1}{V_1} \times 100$$
$$= \frac{1805.23 - 1775.76}{1775.76} \times 100$$
$$= \frac{29.47}{1775.76} \times 100$$
$$= \boxed{1.66\%}$$
$$= \frac{\Delta a}{a_1} \times 100$$

$$
\begin{aligned}
&= \frac{257.89 - 295.96}{295.96} \times 100 \\
&= \frac{-38.07}{295.96} \times 100 \\
&= \boxed{-12.86\%}
\end{aligned}
$$

It is seen that when the supersonic flow is accelerated from Mach 2 to 3, the percentage change in flow speed is 20.26%, and the corresponding decrease in speed of sound is −19.82%. It is explicit that the change in the absolute value of flow speed and sound speed is comparable, that is, $|\Delta V| \approx |\Delta a|$.

When the hypersonic flow is accelerated from Mach 6 to 7, the increase in flow speed and decrease of speed of sound, respectively, are 1.66% and −12.86%. It is seen that the magnitude of the decrease in speed of sound is an order-of-magnitude larger than the increase in the flow speed. This clearly demonstrates that, in the hypersonic regime, change of flow Mach number is dictated by the decrease of speed of sound, and the increase in the flow speed is considerably smaller than the decrease of speed of sound. In other words, it may be stated that, for Mach numbers above 5, the effect of change in the sound speed hypes the change in flow Mach number; flow acceleration or deceleration, and change in flow properties, hence, the flow is aptly called *hypersonic*.

The second question comes to our mind is, whether there is any upper limit for Mach number of hypersonic flow? The answer to this question is *yes*. This limiting Mach number up to which the flow will be hypersonic, and above which the flow will not be amenable to hypersonic theory is about 40, for air at sea level state. Flows with Mach number above 40 would experience very strong temperature effect, leading to dissociation and ionization, and the field is referred to as *high enthalpy gas dynamics*, which encompasses high-temperature and plasma flows.

2.7 Re-entry Problem

For re-entry[2] application, blunt-nosed body is preferable to avoid high stagnation temperature and to achieve efficient heat dissipation. But for sender bodies, one can go up to a very high Mach numbers without taking into account real gas effects. There may be small regions near the leading edge of the body where the kinetic heating is large. Therefore, except at very high freestream Mach number, M_∞, there is no need to consider real gas effects.

For flying machines, such as aeroplane, when treated as slender bodies, real gas effects are small. A re-entry satellite having freestream Mach number $M_\infty \sim 28$

2 For Earth, atmospheric entry occurs by convention at the Karman line at an altitude of 100 km (62 miles; 54 nautical miles) above the surface.

Table 2.1 Supersonic and hypersonic flows.

Supersonic	Hypersonic
Linear equations	Nonlinear equations
Mach waves	Shock waves
Inviscid boundary effect	Viscous interaction
Ideal gas	Real gas

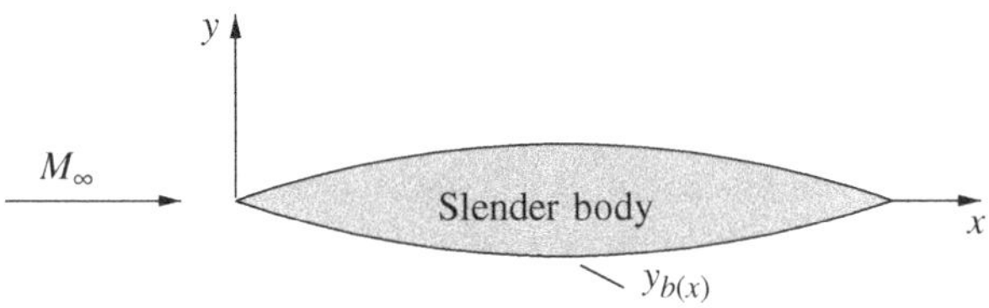

Figure 2.2 A slender body in a uniform flow.

will have to go through hypersonic–supersonic–subsonic regimes. Therefore, we deal only with sharp leading edge with attached shocks. That is, there is no separated (detached) shock. The nature of equations and flow zones associated with supersonic and hypersonic flows are listed in Table 2.1.

2.7.1 Slender Body Geometry

A slender body in a uniform freestream flow is shown in Fig. 2.2. The axis of the body is along the x-direction. The thickness is designated by $y_b(x)$. From the geometry of the body, it can be seen that,

$$\frac{dy_b}{dx} = \theta_b \ll 1$$
$$\tan\,\theta_b \approx \theta_b$$
$$\sin\,\theta_b \approx \theta_b$$
$$\cos\,\theta_b \approx 1$$

2.8 Flow Past a Semi-wedge

Uniform hypersonic flow past a semi-wedge of angle θ is shown in Fig. 2.3. An oblique shock of wave angle β is formed at the nose of the semi-wedge, as shown in the figure. We have $M_\infty \gg 1$; $\theta \ll 1$. If the flow on either side of the shock is assumed to be isentropic, we can express the process relation as

$$\frac{p}{\rho^\gamma} = \text{constant}$$

where p is the static pressure, ρ is the static density, and γ is the isentropic index, given by the ratio of the specific heats, c_p and c_v.

The flow turning angle–shock angle–Mach number (θ-β-M) relation for supersonic flow is [1,12]

$$\frac{\tan(\beta-\theta)}{\tan\beta} = \frac{(\gamma-1)M_1^2\sin^2\beta+2}{(\gamma+1)M_1^2\sin^2\beta}$$

For small flow turning angle, θ, such as the flow past a slender semi-wedge in Fig. 2.3, the shock angle β is also small. For large Mach number, we have $M \gg 1$. Therefore, in the θ–β–M relation, $\tan\theta \approx \theta$, $\cot\beta \approx 1/\beta$, $\sin\beta \approx \beta$, $\cos 2\beta \approx 1$, and the θ-β-M relation simplifies to

$$\theta = \frac{2}{\beta}\left[\frac{M^2\beta^2-1}{M^2(\gamma+1)+2}\right]$$

$$M^2\beta^2 - 1 = \frac{\beta\theta}{2}\left[M^2(\gamma+1)+2\right]$$

$$= \beta\theta\left[\frac{\gamma+1}{2}M^2+1\right]$$

Noting that $\frac{\gamma+1}{2}M^2 \gg 1$, we have

$$M^2\beta^2 - 1 = \beta\,\theta\,\frac{\gamma+1}{2}M^2$$

Diving by $M^2\theta^2$, this becomes

$$\left(\frac{\beta}{\theta}\right)^2 - \frac{1}{M^2\theta^2} = \frac{\gamma+1}{2}\frac{\beta}{\theta}$$

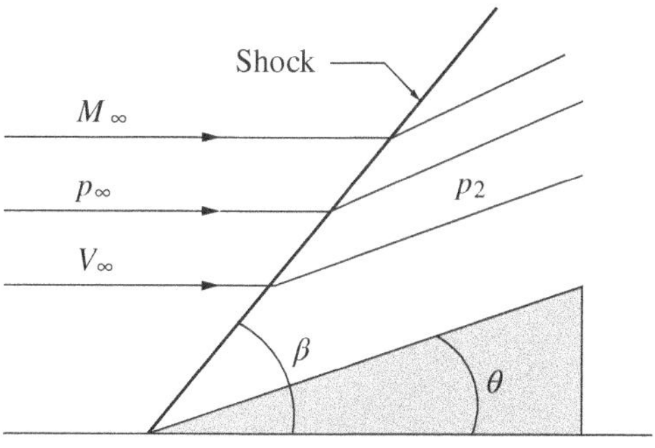

Figure 2.3 Uniform hypersonic flow past a semi-wedge.

or

$$\left(\frac{\beta}{\theta}\right)^2 - \frac{\gamma+1}{2}\frac{\beta}{\theta} - \frac{1}{M^2\theta^2} = 0$$

Solving for β/θ, we get

$$\frac{\beta}{\theta} = \frac{\frac{\gamma+1}{2} + \sqrt{\frac{(\gamma+1)^2}{2^2} + 4\frac{1}{M^2\theta^2}}}{2}$$

$$= \frac{\gamma+1}{4}\left[1 + \sqrt{1 + \frac{\left(\frac{\gamma+1}{4}\right)^2}{(M\theta)^2}}\right]$$

That is,

$$\beta = \frac{\gamma+1}{4}\left[1 + \sqrt{1 + \frac{\left(\frac{\gamma+1}{4}\right)^2}{(M\theta)^2}}\right]\theta \tag{2.10}$$

The Mach number M in Eq. (2.10) is the freestream Mach number M_∞. In the above derivation, the subscript ∞ was dropped for convenience.

From Eq. (2.10), it is seen that the independent parameter is not just M_∞, but $M_\infty\theta$. This term ($M_\infty\theta$) is known as *hypersonic similarity parameter* for slender body configurations.

2.8.1 Pressure Coefficient for Hypersonic Flow

The pressure coefficient, C_p, at any point in a flow is defined as

$$C_p = \frac{p - p_\infty}{\frac{1}{2}\rho_\infty V_\infty^2} \tag{2.11}$$

where p is the static pressure at the point of interest, p_∞, ρ_∞, and V_∞ are the pressure, density, and velocity of the freestream flow, respectively.

The dynamic pressure, $q_\infty = \frac{1}{2}\rho_\infty V_\infty^2$, can expressed in terms of Mach number as

$$q_\infty = \frac{1}{2}\rho_\infty V_\infty^2$$

$$= \frac{1}{2}\frac{p_\infty}{RT_\infty}V_\infty^2$$

Because by state equation, $p_\infty = \rho_\infty R T_\infty$. The sound speed is $a_\infty = \sqrt{\gamma R T_\infty}$. Therefore,

$$\begin{aligned} q_\infty &= \frac{1}{2}\frac{p_\infty}{a_\infty^2/\gamma} \\ &= \frac{\gamma}{2} p_\infty \frac{V_\infty^2}{a_\infty^2} \\ &= \frac{\gamma}{2} p_\infty M_\infty^2 \end{aligned}$$

Therefore, the pressure coefficient becomes

$$\begin{aligned} C_p &= \frac{p - p_\infty}{\frac{\gamma}{2} p_\infty M_\infty^2} \\ &= \frac{2}{\gamma M_\infty^2}\left(\frac{p}{p_\infty} - 1\right) \end{aligned}$$

The pressure coefficient in the zone 2, downstream of the shock at the nose of the semi-wedge in Fig. 2.3, is given by

$$C_p = \frac{2}{\gamma M_\infty^2}\left(\frac{p_2}{p_\infty} - 1\right)$$

From oblique shock theory, the pressure ratio is [1, 12]

$$\frac{p_2}{p_\infty} = 1 + \frac{2\gamma}{\gamma + 1}(M_\infty^2 \sin^2\beta - 1)$$

Therefore,

$$\begin{aligned} C_p &= \frac{2}{\gamma M_\infty^2}\left(\frac{2\gamma}{\gamma + 1}(M_\infty^2 \sin^2\beta - 1)\right) \\ &= \frac{2}{M_\infty^2}\frac{2}{\gamma + 1}\left(M_\infty^2 \sin^2\beta - 1\right) \\ &= \frac{4}{(\gamma + 1)M_\infty^2}\left(M_\infty^2 \sin^2\beta - 1\right) \end{aligned}$$

For $M_\infty^2 \gg 1$, $(M_\infty^2 \sin^2\beta - 1) \approx M_\infty^2 \sin^2\beta$, therefore,

$$C_p = \frac{2}{M_\infty^2}\frac{2}{\gamma + 1} M_\infty^2 \sin^2\beta$$

that is,

$$C_p = \frac{4}{(\gamma + 1)M_\infty^2} M_\infty^2 \sin^2\beta$$

For small shock angles ($\beta \ll 1$), $\sin\beta \approx \beta$ and the C_p becomes

$$C_p = \frac{4}{(\gamma + 1)}\beta^2$$

Substituting the expression for β from Eq. (2.10),

$$C_p = \frac{4}{(\gamma+1)}\left(\frac{\gamma+1}{4}\right)^2\left[1+\sqrt{\frac{\left(\frac{4}{\gamma+1}\right)^2}{M^2\theta^2}}\right]^2\theta^2$$

or

$$\boxed{\frac{C_p}{\theta^2} = \frac{\gamma+1}{4}\left[1+\sqrt{\frac{\left(\frac{4}{\gamma+1}\right)^2}{M^2\theta^2}}\right]^2} \tag{2.12}$$

that is,

$$\boxed{\frac{C_p}{\theta^2} = f_2(M\theta)} \tag{2.12a}$$

At the limit of high Mach number, the expression for the shock angle given by Eq. (2.10) simplifies to

$$\beta = \frac{\gamma+1}{2}\theta \tag{2.13}$$

The density ratio across the shock is [1, 12]

$$\frac{\rho_2}{\rho_1} = \frac{(\gamma+1)M_\infty^2\sin^2\beta}{(\gamma-1)M_\infty^2\sin^2\beta+2}$$

For small β, $\sin\beta \approx \beta$, therefore,

$$\frac{\rho_2}{\rho_1} = \frac{(\gamma+1)M_\infty^2\beta^2}{(\gamma-1)M_\infty^2\beta^2+2}$$
$$= \frac{(\gamma+1)/(\gamma-1)}{1+\frac{2}{\gamma-1}\frac{1}{M_\infty^2\beta^2}}$$

But by Eq. (2.10), β is a function of the hypersonic similarity parameter $(M_\infty\theta)$, thus

$$\frac{\rho_2}{\rho_1} = f_3(M_\infty\theta) \tag{2.14}$$

From Eqs. (2.11) to (2.14), it is seen that $M_\infty\beta$, C_p/θ^2, and ρ_2/ρ_1 are governed only by the hypersonic similarity parameter $(M_\infty\theta)$.

Example 2.7 If the pressure coefficient of a slender body of nose angle 3° in hypersonic flight is 0.0157. If the specific heats ratio of the flow around the body is 1.35, determine the hypersonic similarity parameter.

Solution

Given, $\theta = 3^\circ$, $C_p = 0.0157$, and $\gamma = 1.35$.

By Eq. (2.12), the pressure coefficient is

$$\frac{C_p}{\theta^2} = \frac{\gamma+1}{4}\left[1 + \sqrt{\frac{\left(\frac{4}{\gamma+1}\right)^2}{M^2\theta^2}}\right]^2$$

Therefore,

$$\frac{C_p}{\theta^2} = \frac{1.35+1}{4}\left[1 + \sqrt{\frac{\left(\frac{4}{1.35+1}\right)^2}{M^2\theta^2}}\right]^2$$

$$\frac{0.0157}{(3/57.3)^2} = 0.5875\left[1 + \sqrt{\frac{(4/2.35)^2}{M^2\theta^2}}\right]^2$$

$$= 0.5875\left[1 + \sqrt{\frac{2.897}{M^2\theta^2}}\right]^2$$

$$5.73 = 0.5875\left[1 + \frac{\sqrt{2.897}}{M\theta}\right]^2$$

$$1 + \frac{1.702}{M\theta} = \sqrt{\frac{5.73}{0.5875}}$$

$$1 + \frac{1.702}{M\theta} = 3.123$$

$$2.123\,M\theta = 1.702$$

$$M\theta = \frac{1.702}{2.123}$$

$$= \boxed{0.8}$$

2.9 Hypersonic Limiting Case

In the limiting case, the freestream Mach number of hypersonic flow tends to ∞, that is, $M_\infty \to \infty$, thus $M_\infty\,\theta \to \infty$. Therefore, in the limiting case, the shock angle, β, pressure coefficient C_p/θ^2, and the density ratio ρ_2/ρ_∞ become

$$\beta = \frac{\gamma+1}{2}\theta$$

$$\frac{C_p}{\theta^2} = \gamma + 1$$

$$\frac{\rho_2}{\rho_\infty} = \frac{\gamma + 1}{\gamma - 1}$$

For $\gamma = 1.4$, we have

$$\beta = 1.2\,\theta$$

$$\frac{C_p}{\theta^2} = 2.4$$

$$\frac{\rho_2}{\rho_\infty} = 6$$

Note that the density ratio maximum possible for the case of $M_\infty \to \infty$ is 6. This is the case for supersonic flow also [1, 12].

This limiting case has another limiting case for Newtonian flow with $\gamma = 1$. This flow is extremely simple but is very important. For this case, there is no perturbation in the flow. All fluid elements hit the body and travel tangentially in a straight line.

Note: Newton did this in connection with finding the drag in a low-speed flow. For Newtonian flow: $M_\infty\theta \to \infty, \gamma = 1$. Therefore, $\beta = \theta$ and $C_p = 2\,\theta^2$.

Example 2.8 A slender wedge cruises at Mach 7.2. If the hypersonic similarity parameter is 0.72, determine the nose angle of the wedge.

Solution

Given, $M = 7.2$, and $M\theta = 0.78$.

The hypersonic similarity parameter given by $M\theta$ is 0.78. Therefore, the nose angle θ is

$$\begin{aligned} \theta &= 0.78/M \\ &= 0.78/7.2 \\ &= 0.1083 \ \text{radian} \end{aligned}$$

1 radian = 57.3°, therefore,

$$\begin{aligned} \theta &= 0.78 \times 57.3 \\ &= \boxed{6.2^\circ} \end{aligned}$$

2.10 Newtonian Formula

For the limiting case of Newtonian flow, with the freestream Mach number tending to infinity ($M_\infty \to \infty$) and the specific heats ratio unity ($\gamma = 1$), the shock coincides with the surface of the body present in the flow, as shown in Fig. 2.4. That

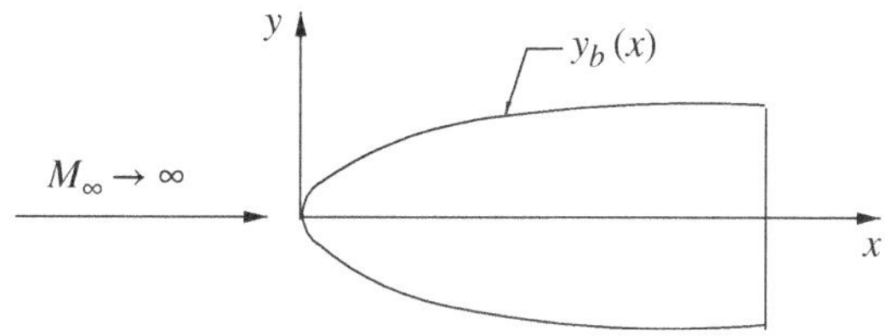

Figure 2.4 Newtonian limiting hypersonic flow past a slender body.

is, the shock is on the body surface itself. Given, a slender body of contour is $y_b(x)$, $\theta_b = \frac{dy_b}{dx}$. Find the pressure distribution over the surface of the body.

Solution: For $M_\infty \to \infty$ and $\gamma = 1$, by Eq. (2.11),

$$C_p = 2\,\theta_b^2(x), \quad \theta \geq 0$$
$$C_p = 0, \quad \theta < 0$$

For the portion behind the maximum thickness location for a body ($dy_b(x)/dx = 0$), such as that shown in Fig. 2.5, only vacuum prevails. This is because

$$\begin{aligned} C_p &= \frac{p_b - p_\infty}{\frac{1}{2}\rho_\infty V_\infty^2} \\ &= \frac{p_b}{\frac{1}{2}\rho_\infty V_\infty^2} - \frac{2}{V_\infty^2}\frac{p_\infty}{\rho_\infty} \\ &= \frac{p_b}{\frac{1}{2}\rho_\infty V_\infty^2} - \frac{2\gamma R T_\infty}{\gamma V_\infty^2} \\ &= \frac{p_b}{\frac{1}{2}\rho_\infty V_\infty^2} - \frac{2a_\infty^2}{\gamma V_\infty^2} \\ &= \frac{p_b}{\frac{1}{2}\rho_\infty V_\infty^2} - \frac{2}{\gamma M_\infty^2} \end{aligned}$$

For $M_\infty \to \infty$,

$$C_p = \frac{p_b}{\frac{1}{2}\rho_\infty V_\infty^2} - 0$$

Therefore, when $p_b = 0$, $C_p = 0$.

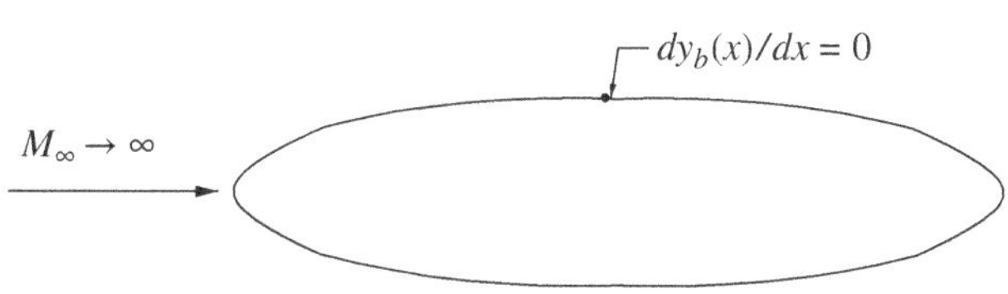

Figure 2.5 A slender body in hypersonic flow.

Example 2.9 A missile cruises at Mach 8 at 10 km altitude. If the base pressure coefficient is −0.223, assuming the air as perfect gas with specific heats ratio 1.4, determine the base pressure.

Solution

Given, $M = 8$, $h = 10$ km, $C_p = -0.223$, and $\gamma = 1.4$. From Table A.2, in the Appendix, at $h = 10,000$ m, $\rho = 0.413$ kg/m^3, and $T = 223.15$ K.

The speed of sound is

$$\begin{aligned} a &= \sqrt{\gamma R T} \\ &= \sqrt{1.4 \times 287 \times 223.15} \\ &= 299.44\,\text{m/s} \end{aligned}$$

The velocity of the missile is

$$\begin{aligned} V &= M a \\ &= 8 \times 299.44 \\ &= 2395.52\,\text{m/s} \end{aligned}$$

By Eq. (2.11), the pressure coefficient can be expressed as

$$C_p = \frac{p_b}{\frac{1}{2}\rho V^2} - \frac{2}{\gamma M^2}$$

where p_b is the pressure at the base of the missile. Therefore,

$$\begin{aligned} \frac{p_b}{\frac{1}{2}\rho V^2} &= C_p + \frac{2}{\gamma M^2} \\ &= -0.233 + \frac{2}{1.4 \times 8^2} \\ &= -0.211 \\ p_b &= -0.211 \times \frac{1}{2}\rho V^2 \\ &= -0.211 \times \frac{1}{2} \times 0.413 \times 2395.52^2 \\ &= \boxed{-250\,\text{kPa (gauge)}} \end{aligned}$$

2.11 Surface Pressure Distribution

The pressure distribution over the surface of the slender body in a hypersonic stream will be as shown in Fig. 2.6.

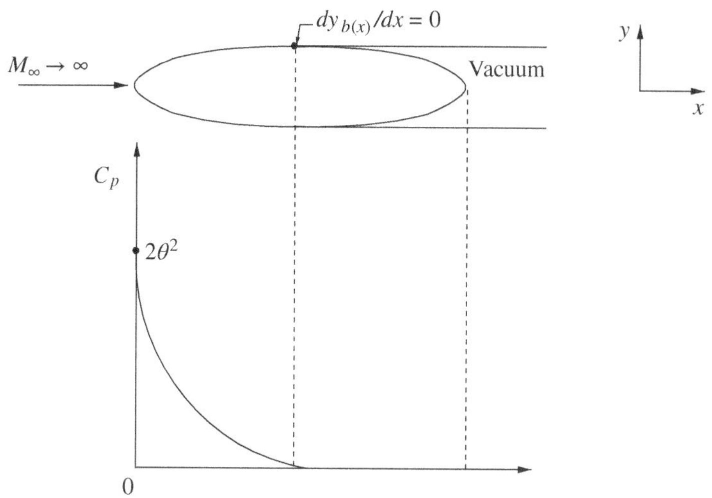

Figure 2.6 Pressure distribution over the surface of a slender body in hypersonic flow.

2.12 Modified Newtonian Formula

To take into account the effect of the freestream Mach number, M_∞, empirically, the Newtons formula depends on geometry only. At the leading edge of the body ($x = 0$), the pressure coefficient C_p is equal to that given by the oblique shock, and later on (i.e. downstream of the leading edge) the dependence is as per Newtonian formula,

$$C_p = F \cdot \frac{\theta_b(x)^2}{\theta_0^2}$$

where F is the pressure coefficient given by Eq. (2.11)

$$F = \frac{C_p}{\theta^2} = \frac{\gamma+1}{4}\left[1 + \sqrt{\frac{\left(\frac{4}{\gamma+1}\right)^2}{M^2\theta^2}}\right]\theta_0^2 \tag{2.15}$$

The similarity parameter is

$$\boxed{K = M_\infty \theta_0} \tag{2.16}$$

This K is known as *hypersonic similarity parameter*.

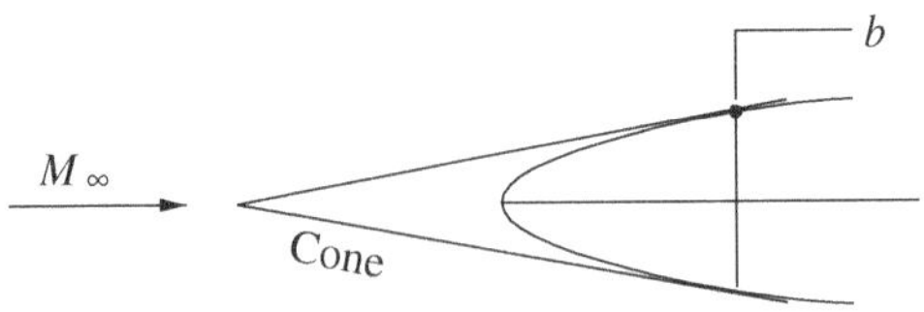

Figure 2.7 Tangent cone.

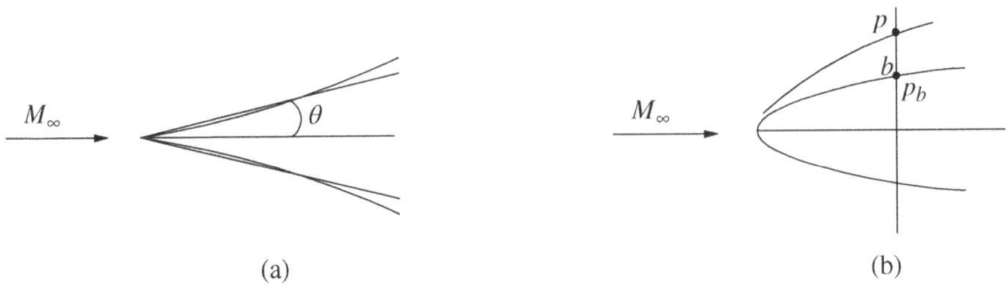

Figure 2.8 (a) Convex and (b) concave cones in hypersonic flow.

2.13 Tangent Wedge or Tangent Cone Method

Here the distribution is similar to the modified Newtonian formula, excepting for taking $\theta(x)$ instead of θ_0. The pressure distribution at the point b on the cone pressure distribution is the same as that due to a cone or a wedge which has slope the same as the inclination (local inclination). Hence, tangent cone or tangent wedge, as shown in Fig. 2.7, can be taken as identical shapes experiencing the same pressure distribution.

It is essential to note that the Newtonian theory is not good for some shapes. Experiments show that there is an appreciable difference in the pressure distribution predicted by the Newtonian model and the actual one. For both concave and convex body, as shown in Fig. 2.8, the pressure at point p outside the body inline with the point b on the body (Fig. 2.8(b)) is not the same because of centrifugal force.

2.14 Busemann Correction for Centrifugal Force

Busemann applied correction for the pressure distribution by considering the centrifugal force. The centrifugal force acting on the cone, as shown in Fig. 2.9, is

$$\frac{\partial p}{\partial n} = \frac{\rho V^2}{R_c}$$

Figure 2.9 Concave cone in hypersonic flow.

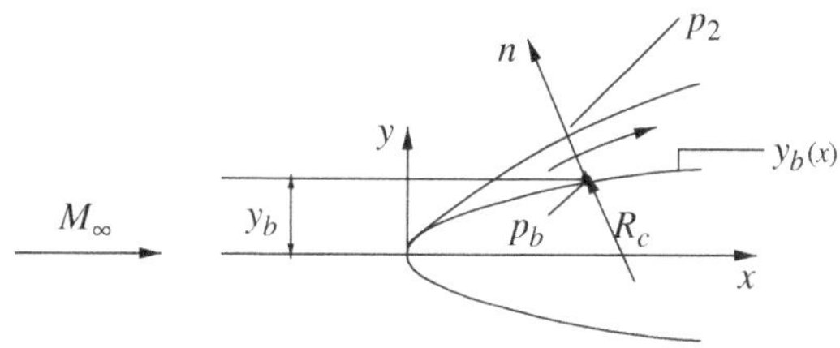

where p is the pressure, n is the direction normal to the body surface, ρ is the flow density, V is the flow velocity, and R_c is the radius of curvature of the body.

The pressure difference between the point outside the cone and on the cone surface is

$$p_2 - p_b = \int_0^\delta \frac{\rho V^2}{R_c} dn$$
$$= y_b'' \overline{V} \int_0^\delta \rho V \, dn$$

The radius of curvature is given by

$$\frac{1}{R_c} = \frac{y_b''}{(1 + y'^2)^{3/2}}$$
$$\approx y_b'' \overline{V} \int_0^\delta \rho V \, dn$$

Therefore,

$$p_2 - p_b = \rho_\infty V_\infty^2 y_b y_b''$$

For the Newtonian flow, the pressure coefficient is

$$C_p = \frac{p_b - p_\infty}{\frac{1}{2}\rho_\infty V_\infty^2}$$
$$= \frac{p_b}{\frac{1}{2}\rho_\infty V_\infty^2}$$

or

$$C_p = 2\theta^2 \tag{2.17}$$

Therefore, for two-dimensional bodies

$$C_p = \frac{p_b - p_\infty}{\frac{1}{2}\rho_\infty V_\infty^2}$$
$$= 2 y_b y_b''$$

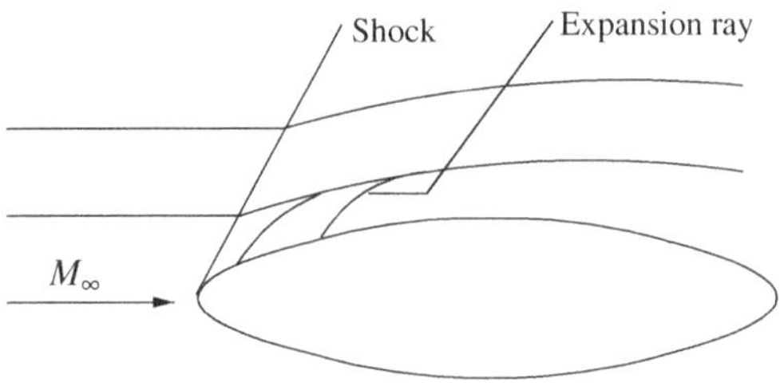

Figure 2.10 Hypersonic flow past a slender body.

or

$$C_p = \frac{p_b - p_\infty}{\frac{1}{2}\rho_\infty V_\infty^2} = 2R_b R_b''$$

All these methods give pressures only up to the point where $y_b' = 0$. After that point, there is a vacuum. But in reality, there will be no vacuum. This pressure can be obtained from shock-expansion theory.

In the actual flow, there is an expansion, as shown in Fig. 2.10. In the simple shock-expansion theory, the reflection of the expansion waves from the shock is reflected. This theory is simple compared to the actual flow.

Example 2.10 A slender body cruises with freestream density 0.8 kg/m^3. If the pressure gradient normal to the surface at the local point where the radius of curvature is 1.2 m is 0.4 MPa/m^3, determine the cruising speed.

Solution

Given, $\rho = 0.8$ kg/m^3, and $R_c = 1.2$ m. The centrifugal force acting on the body is given by

$$\frac{\partial p}{\partial n} = \frac{\rho V^2}{R_c}$$

Therefore,

$$0.4 \times 10^6 = \frac{0.8 \times V^2}{1.2}$$

$$V^2 = \frac{1.2 \times 0.4 \times 10^6}{0.8} = 0.6 \times 10^6$$

$$V = \sqrt{0.6 \times 10^6} = \boxed{774.6\,\text{m/s}}$$

2.15 Shock-Expansion Method

For more accurate analysis we need to use the method of characteristics. Consider the Prandtl–Meyer expansion of a flow of Mach number M_∞, turning through θ, as shown in Fig. 2.11.

Let the specific heats ratio of the gas be γ. For this expanding flow, the Mach number M after the expansion is greater than the freestream Mach number, $M > M_\infty$, and the flow process is termed as Prandtl–Meyer expansion. For this process, the Prandtl–Meyer function, ν, is the governing parameter, given by [1, 12],

$$\nu = \sqrt{\frac{\gamma+1}{\gamma-1}}\left[\tan^{-1}\sqrt{\frac{\gamma-1}{\gamma+1}(M^2-1)} - \tan^{-1}\sqrt{\frac{\gamma-1}{\gamma+1}(M_\infty^2-1)}\right]$$
$$-\tan^{-1}\sqrt{M_1^2-1} - \tan^{-1}\sqrt{M_\infty^2-1} \tag{2.18}$$

Prandtl–Meyer function ν can be simplified if $M_\infty^2 \gg 1$ and $M \gg 1$. The group $(\tan^{-1}(u) - \tan^{-1}(v))$ can be expressed as

$$\begin{aligned}\tan^{-1}(u) - \tan^{-1}(v) &= \tan^{-1}\frac{u-v}{1+uv}\\ &= \left(1-\frac{v}{u}\right)\frac{1}{v}\end{aligned}$$

The perturbation velocity components u and v are large, therefore,

$$\nu = \sqrt{\frac{\gamma+1}{\gamma-1}}\left(1-\frac{M_\infty}{M}\right)\sqrt{\frac{\gamma+1}{\gamma-1}}\frac{1}{M_\infty} - \left(1-\frac{M_\infty}{M}\right)\frac{1}{M_\infty}$$

Also,

$$M_\infty\theta = f(\gamma, M\nu)$$

These lead to the following relations for M_∞/M and p_∞/p, in terms of the hypersonic similarity factor $(M_\infty\theta)$.

$$\boxed{\frac{M_\infty}{M} = 1 - \frac{\gamma-1}{2}(M_\infty\theta)} \tag{2.19}$$

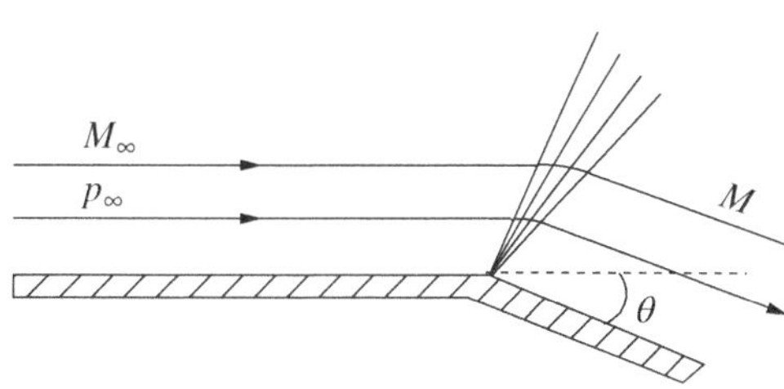

Figure 2.11 Expanding flow.

$$\boxed{\frac{p_\infty}{p} = \left(\frac{M_\infty}{M}\right)^{2\gamma/(\gamma-1)} = \left[1 - \frac{\gamma-1}{2}(M_\infty \theta)\right]^{2\gamma/(\gamma-1)}} \tag{2.20}$$

The pressure coefficient becomes

$$\begin{aligned} C_p &= \frac{p - p_\infty}{\frac{1}{2}\rho_\infty V_\infty^2} \\ &= \left(\frac{p}{p_\infty} - 1\right)\frac{2}{\gamma M_\infty^2} \end{aligned}$$

For a more general case where ν is changing from point to point, the flow turning angle θ in the expressions for M_∞/M and p_∞/p, in Eqs. (2.19) and (2.20), should be taken as a variable.

Example 2.11 A Mach 4 airflow passes through a 6° expansion corner. Determine the Mach number and pressure coefficient downstream of the expansion corner.

Solution

Given, $M_\infty = 4$, and $\theta = 6°$.

By Eq. (2.19),

$$\frac{M_\infty}{M} = 1 - \frac{\gamma - 1}{2} M_\infty \theta$$

Therefore,

$$\begin{aligned} \frac{M_\infty}{M} &= 1 - \frac{1.4 - 1}{2}\left(4 \times \frac{6}{57.3}\right) \\ &= 1 - 0.2 \times 0.4188 \\ &= 0.9162 \end{aligned}$$

$$\begin{aligned} M &= \frac{M_\infty}{0.9162} \\ &= \frac{4}{0.9162} \\ &= \boxed{4.37} \end{aligned}$$

The pressure coefficient is given by

$$C_p = \left(\frac{p}{p_\infty} - 1\right)\frac{2}{\gamma M_\infty^2}$$

But, by Eq. (2.20),

$$\frac{p}{p_\infty} = \left(\frac{M_\infty}{M}\right)^{2\gamma/(\gamma-1)}$$

$$= \left(\frac{4}{4.37}\right)^{(2\times1.4)/(1.4-1)}$$
$$= 0.538$$

Therefore,

$$C_p = \left(\frac{1}{0.538} - 1\right) \times \frac{2}{1.4 \times 4^2}$$
$$= (1.859 - 1) \times 0.0893$$
$$= \boxed{0.0767}$$

Aliter:

For $M_\infty = 4$, from isentropic table [1], the Prandtl–Meyer function is $\nu_\infty = 65.785°$.

The Prandtl–Meyer function after the expansion fan is

$$\nu = 65.785 + 4 = 69.785°$$

Now, from isentropic table, for $\nu = 69.785°$,

$$M = \boxed{4.32}$$

Example 2.12 A body is cruising at Mach 6. If the pressure coefficient at a point on its surface is −0.032, find the Mach number at that point, assuming the specific heats ratio as 1.4.

Solution

Given, $M_\infty = 6$, $C_p = -0.032$, and $\gamma = 1.4$.

The pressure coefficient at any point on the body surface is given by

$$C_p = \left(\frac{p}{p_\infty} - 1\right) \frac{2}{\gamma M_\infty^2}$$

Therefore,

$$-0.032 = \left(\frac{p}{p_\infty} - 1\right) \times \frac{2}{1.4 \times 6^2}$$
$$\frac{p}{p_\infty} - 1 = \frac{-0.032 \times 1.4 \times 6^2}{2}$$
$$= -0.8064$$
$$\frac{p}{p_\infty} = 0.1936$$
$$\frac{p}{p_0} \times \frac{p_0}{p_\infty} = 0.1936$$

For $M_\infty = 6$, from isentropic table, $\frac{p_\infty}{p_0} = 0.6333 \times 10^{-3}$. Therefore,

$$\frac{p}{p_0} = 0.1936 \times 0.6333 \times 10^{-3}$$

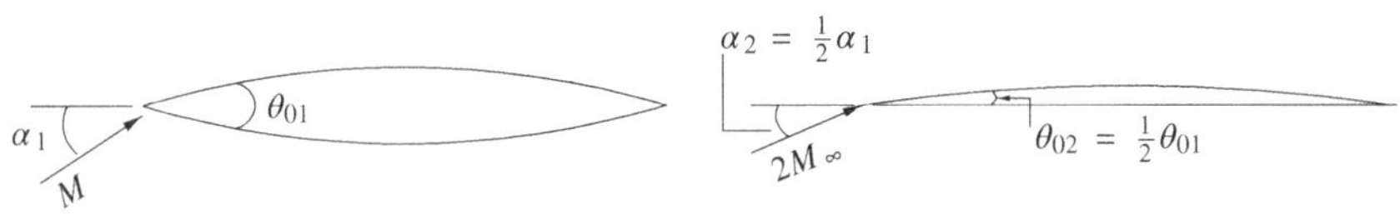

Figure 2.12 Similar flows.

$$= 0.1226 \times 10^{-3}$$

For $\frac{p}{p_0} = 0.1226 \times 10^{-3}$, from isentropic table,

$$M = \boxed{7.8}$$

2.15.1 For Similar Flow

For similar flows such as those shown in Fig. 2.12, the optimization is done through the geometrical optimization. This kind of bodies is known as power-law bodies.

2.15.2 For Slender Bodies with Minimum Pressure-Drag in Symmetrical Flow in the Hypersonic Flow Region

For two-dimensional bodies of shape given by $y_b \sim x^n$, the drag is given as

$$C_D = \frac{D}{q_\infty b\,\delta}$$

where $\delta = d/c$, d is the maximum thickness, b is the span, and c is the chord.

Drag coefficient given by Newton's model and Newton–Busemann model for two-dimensional and axisymmetric bodies are listed in Tables 2.2 and 2.3.

Example 2.13 If a two-dimensional slender body of span 3 m, chord 1 m, and the maximum thickness as 10% of the chord flies at freestream pressure 500 mm of mercury experiences 27 kN drag, assuming the flow as Newtonian and symmetrical, with specific heats ratio 1.4, find the flight Mach number.

Solution

Given, $b = 3$ m, $c = 1$ m, $\delta = 0.1$ m, $\gamma = 1.4$, and $D = 27$ kN.

The freestream pressure is

$$p = \frac{500}{760} \times 101\,325$$

$$= 66\,661 \text{ Pa}$$

The drag is given by

$$D = q\,S\,C_D$$

Table 2.2 Drag for two-dimensional bodies of shape, $y \sim x^n$ and $C_D = D/(q_\infty b\,d)$.

Given	n	Drag (Newton)	n	Drag (Newton–Busemann)
d, c	1.0	$\frac{C_D}{\delta^2} = 0.5$	0.864	$\frac{C_D}{\delta^2} = 0.459$
Axi-symmetric d, y	1.5	$C_D\left(\frac{y}{b\,d^2}\right)^2 = 0.108$	1.000	$C_D\left(\frac{y}{b\,d^2}\right)^2 = 0.125$
d, O	1.0	$C_D\left(\frac{O}{b\,d}\right)^2 = 2$	0.864	$C_D\left(\frac{O}{b\,d}\right)^2 = 1.830$

Table 2.3 Drag for axisymmetric bodies of shape, $R_b \sim x^n$, $\delta = d/c = R_{b_{max}}/c$, and $C_D = D/(q_\infty R^2_{b_{max}})$.

Given	n	Drag (Newton)	n	Drag (Newton–Busemann)
d, c	0.75	$\frac{C_D}{\delta^2} = 0.422$	0.864	$\frac{C_D}{\delta^2} = 0.459$
d, V	1.5	$C_D\left(\frac{V}{R^3_{b_{max}}}\right)^2 = 2.08$	0.876	$C_D\left(\frac{V}{R^2_{b_{max}}}\right)^2 = 2.166$
d, O	1.0	$C_D\left(\frac{O}{R^2_{b_{max}}}\right)^2 = 2$	0.761	$C_D\left(\frac{O}{\pi\,R^2_{b_{max}}}\right)^2 = 1.840$

where

$$
\begin{aligned}
q &= \frac{1}{2}\rho V^2 \\
&= \frac{1}{2}\frac{p}{RT}V^2 \\
&= \frac{1}{2}\frac{\gamma p}{\gamma R T}V^2 \\
&= \frac{1}{2}\frac{\gamma p}{a^2}V^2 \\
&= \frac{\gamma p}{2}M^2
\end{aligned}
$$

Substituting this, the drag expression becomes

$$D = \frac{\gamma p M^2}{2} S C_D$$

For the given shape and flying state, from Table 2.2,

$$
\begin{aligned}
\frac{C_D}{\delta^2} &= 0.5 \\
&= 0.5\,\delta^2 \\
C_D &= 0.5 \times 0.1^2 \\
&= 0.005
\end{aligned}
$$

Therefore,

$$D = \frac{1.4 \times 66\,661 \times M^2}{2} \times (3 \times 1) \times 0.005$$

$$\begin{aligned} M^2 &= \frac{2D}{1.4 \times 66\,661 \times 3 \times 0.005} \\ &= \frac{2 \times 27\,000}{1.4 \times 66\,661 \times 3 \times 0.005} \\ &= 38.575 \\ &= \boxed{6.211} \end{aligned}$$

Example 2.14 A two-dimensional slender body of chord 3 m and maximum thickness 0.4 m has a shape given by $y = x^n$. If $n = 0.75$, determine the drag coefficient using Newton's model theory.

Solution

Given, $c = 3$ m and $d = 0.4$ m. Therefore,

$$\begin{aligned} \delta &= \frac{d}{c} \\ &= \frac{0.4}{3} \\ &= 0.133 \end{aligned}$$

From Table 2.2, for this model, $C_D/\delta^2 = 0.5$. Therefore,

$$\begin{aligned} C_D &= 0.5\,\delta^2 \\ &= 0.5 \times 0.133^2 \\ &= \boxed{0.0088} \end{aligned}$$

2.16 Theory of Slender Hypersonic Bodies

This theory gives exact solution. It would capture the waves prevailing over the body, along with expansion wave reflection, etc.

2.16.1 Basic Equations

Consider the two-dimensional slender body in a hypersonic stream of Mach number M_∞, as shown in Fig. 2.13.

For the case of slender geometry, the three-dimensional flow theory is not much different from the two-dimensional case. For this body, the perturbations caused by the body presence can be regarded as small. Therefore, the velocity along the

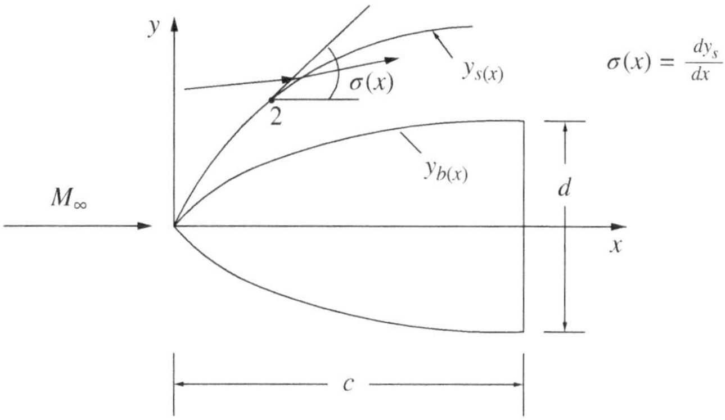

Figure 2.13 A slender two-dimensional body in a hypersonic flow.

x- and y-directions are $(V_\infty + u)$ and v, respectively. Also, $u \ll V_\infty$ and $v \ll V_\infty$. The governing equations for this case are the following.

Continuity equation,

$$\frac{\partial}{\partial x}\left[\rho(V_\infty + u)\right] + \frac{\partial}{\partial y}\left[\rho v\right] = 0$$

Momentum equation,

$$(V_\infty + u)\frac{\partial u}{\partial x} + v\frac{\partial u}{\partial y} + \frac{1}{\rho}\frac{\partial p}{\partial x} = 0$$

$$(V_\infty + u)\frac{\partial v}{\partial x} + v\frac{\partial v}{\partial y} + \frac{1}{\rho}\frac{\partial p}{\partial y} = 0$$

Using the above three equations, four unknowns: u, v, p, and ρ to be solved. Therefore, one more equation is required to find solution for the problem under consideration. It is known that the entropy is constant along the streamlines, therefore,

$$\left(V_\infty + u\right)\frac{\partial}{\partial x}\left(\frac{p}{\rho^\gamma}\right) + v\frac{\partial}{\partial y}\left(\frac{p}{\rho^\gamma}\right) = 0 \tag{2.21}$$

where p/ρ^γ = constant is the process equation.

Using these continuity, momentum, and the process equations, the unknowns $u(x, y)$, $v(x, y)$, $p(x, y)$, and $\rho(x, y)$ can be solved.

If $\sigma(x)$ is the slope of the shock, considering a point 2 just behind the shock, as shown in Fig. 2.13, the boundary conditions on the shock can be expressed as follows.

Boundary conditions at the shock are

$$u_2 = -\frac{2V_\infty}{\gamma + 1}\left(1 - \frac{1}{M_\infty^2\sigma^2}\right)\sigma^2$$

$$v_2 = \frac{2V_\infty}{\gamma+1}\left(1 - \frac{1}{M_\infty^2\sigma^2}\right)\sigma$$

$$p_2 = \rho_\infty 2V_\infty^2\left[\frac{1}{\gamma M_\infty^2\sigma^2} + \frac{2}{\gamma+1}\left(1 - \frac{1}{M_\infty^2\sigma^2}\right)\right]\sigma^2$$

$$\rho_2 = \frac{\gamma+1}{\gamma-1}\rho_\infty\left[1 - \frac{2}{\gamma-1}\frac{1}{M_\infty^2\sigma^2}\right]^{-1}$$

The boundary condition on the body (kinematic flow condition) is

$$\theta_b = \frac{dy_b}{dx} = \frac{v}{V_\infty + u}$$

2.16.2 New Coordinate System

Let

$$u = V_\infty\,\delta^2\,u'$$
$$v = V_\infty\,\delta^2\,v'$$
$$p = \rho_\infty V_\infty^2\,\delta^2\,p'$$
$$\rho = \rho_\infty\,\rho'$$
$$x = cx'$$
$$y = \delta_b y'$$
$$\sigma = O(\delta)$$

where $\delta = d/c$.

With these coordinate systems, the governing equations can be simplified as follows.

Continuity equation:

$$\frac{\partial\rho'}{\partial x'} + \frac{\partial(\rho' v')}{\partial y'} = 0$$

The first equation of motion is not required due to slenderness.

Second equation of motion:

$$\frac{\partial v'}{\partial x'} + v'\frac{\partial v'}{\partial y'} + \frac{1}{\rho'}\frac{\partial p'}{\partial y'} = 0$$

The entropy change relation:

$$\frac{\partial}{\partial x'}\left(\frac{p'}{\rho'^{\gamma}}\right) + v'\frac{\partial}{\partial y'}\left(\frac{p'}{\rho'^{\gamma}}\right) = 0$$

Using these three equations, the three unknowns, v', p', and ρ', can be solved.

2.16.3 Boundary Conditions

The velocity, pressure, and density behind the shock are given by

$$v_2' = \frac{2}{\gamma+1}\left[1 - \left(\frac{\delta}{\sigma}\right)^2 \frac{1}{k^2}\right]\frac{\sigma}{\delta}$$

$$p_2' = \frac{2}{\gamma+1}\left[1 - \frac{\gamma-1}{2\gamma}\left(\frac{\delta}{\sigma}\right)^2 \frac{1}{k^2}\right]\left(\frac{\sigma}{\delta}\right)^2$$

$$\rho' = \frac{\gamma+1}{\gamma-1}\left[1 + \frac{2}{\gamma-1}\left(\frac{\delta}{\sigma}\right)^2 \frac{1}{k^2}\right]^{-1}$$

On the body surface

$$v' = \frac{v_b}{\delta}$$

With these, the solution is of the form

$$v'(x', y' k, \gamma)$$
$$p'(x', y' k, \gamma)$$
$$\rho'(x', y' k, \gamma)$$

This gives the similarity rule for hypersonic flow, that is, for the same k, the flows will be similar whatever may be the freestream Mach number M_∞ and δ. That is, the solution for the flow depends only on the hypersonic similarity parameter,

$$\boxed{k = M_\infty \delta}$$

But $y' = y'(x)$, therefore,

$$v'(x', k, \gamma)$$
$$p'(x', k, \gamma)$$
$$\rho'(x', k, \gamma)$$

The pressure coefficient becomes

$$C_p = \frac{p - p_\infty}{\frac{1}{2}\rho_\infty V_\infty^2}$$
$$= \delta^2 F(x', k, \gamma)$$

This gives rise to another form of similarity rule:

$$\boxed{\frac{C_p}{\delta^2} = F\left(\frac{x}{c}, k, \gamma\right)} \tag{2.22}$$

2.17 Principle of Equivalence for Slender Hypersonic Bodies

In the equations of motion (together with continuity equation), if x' is replaced by time t, the solutions for one-dimensional unsteady case can be obtained. That is, one-dimensional unsteady flow can be solved from equations of motion of the two-dimensional steady flow.

Hypersonic flow past a slender body which is expanding with time, as shown in Fig. 2.14, can be analyzed with the above procedure.

Exact solutions can be obtained for the following cases:

i) for the case of constant ρ, v, and p, trivial solution for wedge flow can be obtained.
ii) for a conical flow, not trivial but exact solution can be obtained.
iii) for the case of similarity parameter tending to infinity ($k \to \infty$), the boundary conditions become very simple.

For $y_b = x'' \,:\, k \to \infty$. For this case, as shown in Fig. 2.15, it turns out that the flow is similar at any x. That is, now there is only one parameter and not two and the flow at any station can be obtained if the flow at that station is known.

$$y_b \sim e^{ax}$$

In this case, the equations are not partial differential equations.

Shock position and pressure distribution over the surface of a slender two-dimensional power law bodies (similar solutions, $M_\infty = \infty$) are shown in Fig. 2.16 and 2.17, respectively.

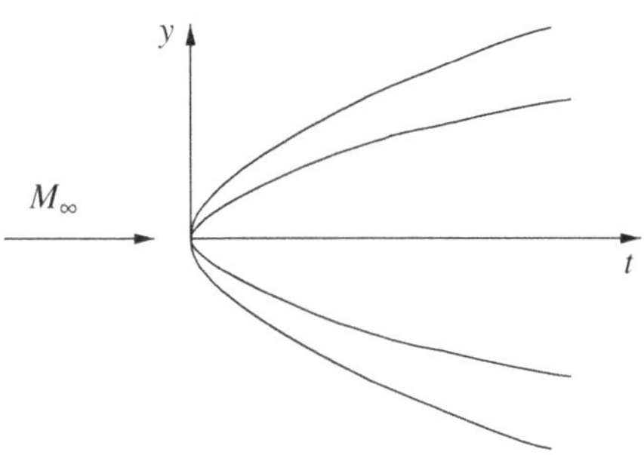

Figure 2.14 A body expanding with time.

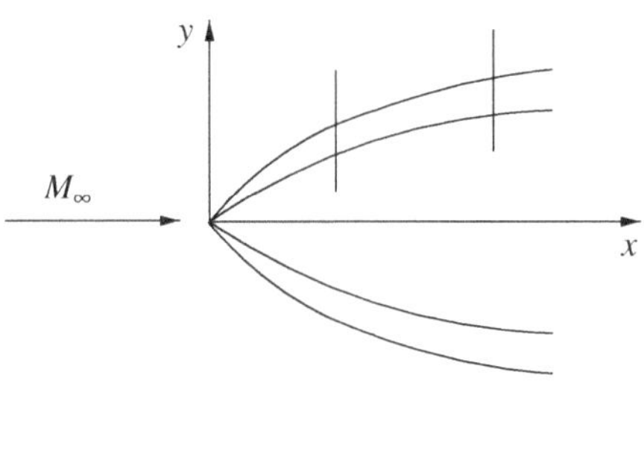

Figure 2.15 A slender body in hypersonic flow.

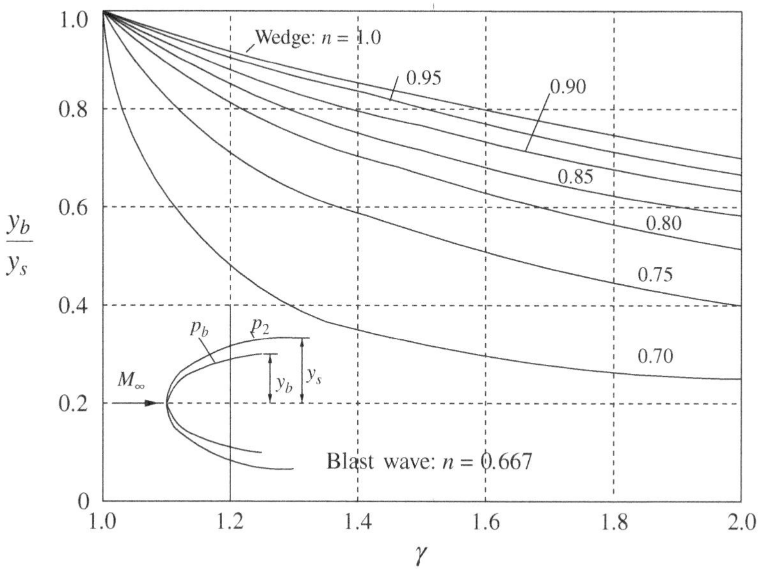

Figure 2.16 Shock position for slender two-dimensional body in a hypersonic flow, $M_\infty \to \infty$.

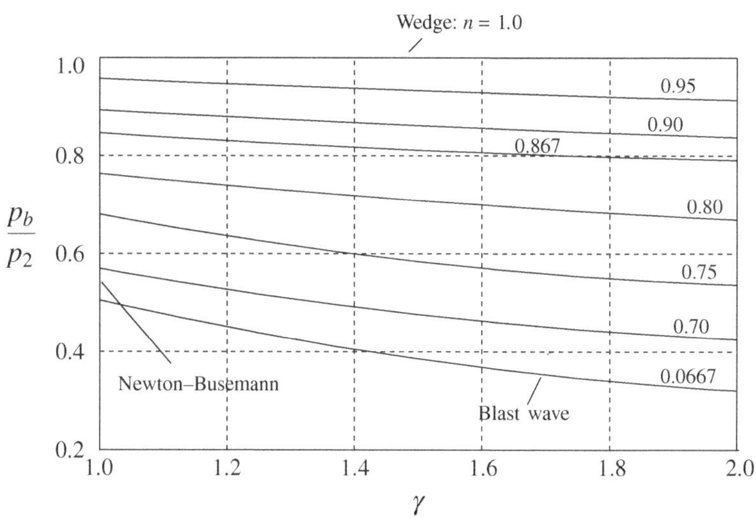

Figure 2.17 Pressure distribution for slender two-dimensional body in a hypersonic flow.

$n = 0.667$: no body, the body thickness = 0
$n = 1$: wedge.
$\gamma = 1$: Newton–Busemann flow. Exact solution.

2.18 Design of Three-Dimensional Hypersonic Slender Bodies

The wave-rider, as shown in Fig. 2.18, is a typical three-dimensional slender body.

A wave-rider is a hypersonic aircraft design that improves its supersonic lift-to-drag ratio by using the shock waves being generated by its own flight as a lifting surface (i.e. treating the waves coinciding with its surface), a phenomenon known as *compression lift*.[3] It consists of a delta-wing planform with a low wing loading to provide considerable surface area to dump the heat generated during its reentry flight phase.

2.18.1 Cone Flow Wave-Rider

As the study of hypersonic flows improved, researchers were able to study wave-rider designs that used different shock wave shapes, the simplest being the conical shock generated by a cone, as illustrated in Fig. 2.19. In these cases, a wave-rider is designed to keep the surrounded shock wave attached to its wings, not a flat sheet, which increases the volume of air trapped under the surface, and thereby increases the lift [13, 14].

For a conical wave-rider with the body shape given by $y_b \sim e^{ax}$, varying shock pattern and the optimum shape can be obtained by varying the constant a in the body expression [13].

How to cut out new bodies from known fields was explained by Seddon and Spence [15] and Jones et al. [16]. The wave-rider has the shape only for the flow corresponding to the design condition.

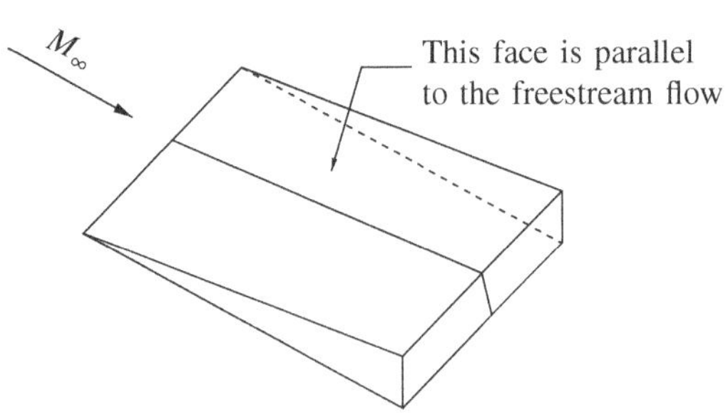

Figure 2.18 A wave-rider in hypersonic flow.

3 In aerodynamics, compression lift refers to an aircraft that uses shock waves generated by its own supersonic flight to generate lift.

Figure 2.19 Cone flow wave-rider.

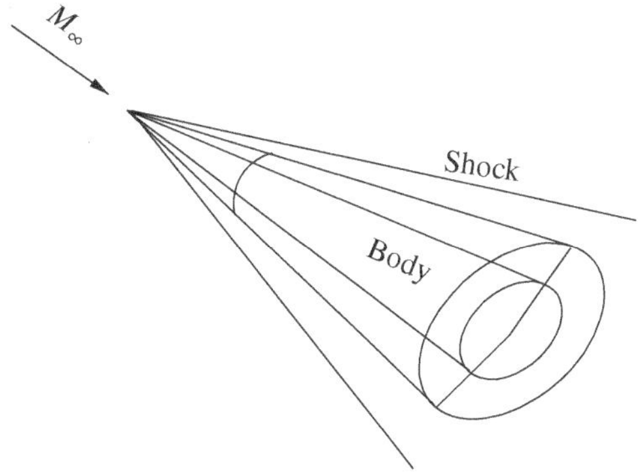

2.19 Caret-Wing or Nonweiler Wing or Wave-Rider

Schematic drawing of a typical caret or Nonweiler wing[4] is shown in Fig. 2.20. This kind of wave-rider could be obtained from such a wedge flow or a cone flow or in general from the power law flow.

Figure 2.20 Nonweiler's wing.

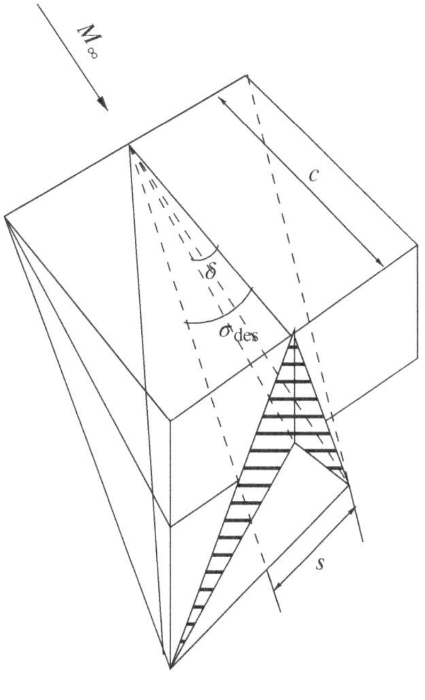

4 In the 1950s, Terrence Nonweiler produced a design for a delta-winged aircraft with a deeply concave underside known as a "Caret-wing."

The geometry is defined by δ, σ_{des}, $s/c = A/4$.

$$\sigma_{\text{des}} = \frac{\gamma+1}{4}\left[1+\sqrt{1+\frac{\left(\frac{4}{\gamma+1}\right)^2}{(M_{\text{des}}\delta)^2}}\right]\delta \tag{2.23}$$

M_{des} also can be considered as a geometric parameter. The pressure coefficient is given by

$$C_{pl} = \frac{p-p_\infty}{\frac{1}{2}\rho_\infty V_\infty^2} = 2\,\sigma_{\text{des}}\,\delta \tag{2.24}$$

The pressure coefficient can also be expressed as

$$\frac{C_{pl}}{\delta^2} = f(\gamma, M_{\text{des}}\delta)$$

The lift and drag coefficients are given by

$$C_L = C_p$$
$$C_D = C_p\,\delta$$

The aerodynamic efficiency becomes

$$\frac{L}{D} = \frac{1}{\delta}$$

Geometry:

Sometimes other parameters are used for geometry. Some of the popular ones are the following.

2.19.1 Volume Parameter

The volume parameter, τ, is defined as

$$\tau = \frac{\mathbf{V}}{S^{3/2}} = \frac{2}{3}\frac{\delta}{\sqrt{A}} \tag{2.25}$$

So, the other possibilities of representing geometry are:

$$\delta\ \sigma_{\text{des}}\ A$$
$$\delta\ M_{\text{des}}\ A$$
$$\tau\ M_{\text{des}}\ A$$

These are commonly used. These solutions are exact, i.e. no approximations are made here except that it is slender.

Sometimes the approximate Newtonian formula, given below, is used.

Approximate Newtonian Formula

The pressure coefficient is

$$C_{pl} = 2\,\delta_{\text{des}}^2$$

Now everything is known about the design condition of caret-wing or wave-riders.

2.20 Off-design Conditions for Caret-Wings

2.20.1 Hypersonic Slender Body Theory

For the caret-wing, as shown in Fig. 2.21, the pressure coefficient is given by

$$\frac{C_p}{\delta^2} = F\left(k,\, \frac{\alpha}{\delta},\, A\,M_{\text{des}},\, \tau\sqrt{M_{\text{des}}},\, \gamma,\, \frac{y}{s}\right) \tag{2.26}$$

Typical shock shapes of caret-wings at $\alpha = 0°$ for various Mach numbers are shown in Fig. 2.22.

In Fig. 2.22, M_{det} is the shock detachment Mach number. When $M_\infty > M_{\text{des}}$, the cross flow component is more and so the shock is pushed nearer to the body.

Shock detachment Mach number, M_{det}, is the freestream Mach number M_∞ at which the shock gets just detached from the leading edge. In Fig. 2.22(a)–(d), the angle of incidence α of the caret-wing is kept constant, and the freestream Mach number M_∞ is changed. Instead, one could keep the M_∞ constant and change α. An increase in M_∞ with constant α is equivalent to keeping M_∞ constant and increasing α. This can be seen from the expression for pressure coefficient,

$$\frac{C_p}{\delta^2} = F\left(k,\, \frac{\alpha}{\delta},\, A\,M_{\text{des}},\, \tau\sqrt{M_{\text{des}}},\, \gamma,\, \frac{y}{s}\right)$$

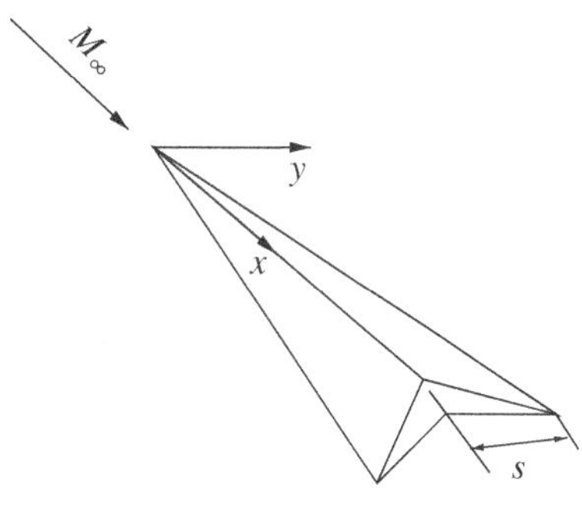

Figure 2.21 Nonweiler's wing.

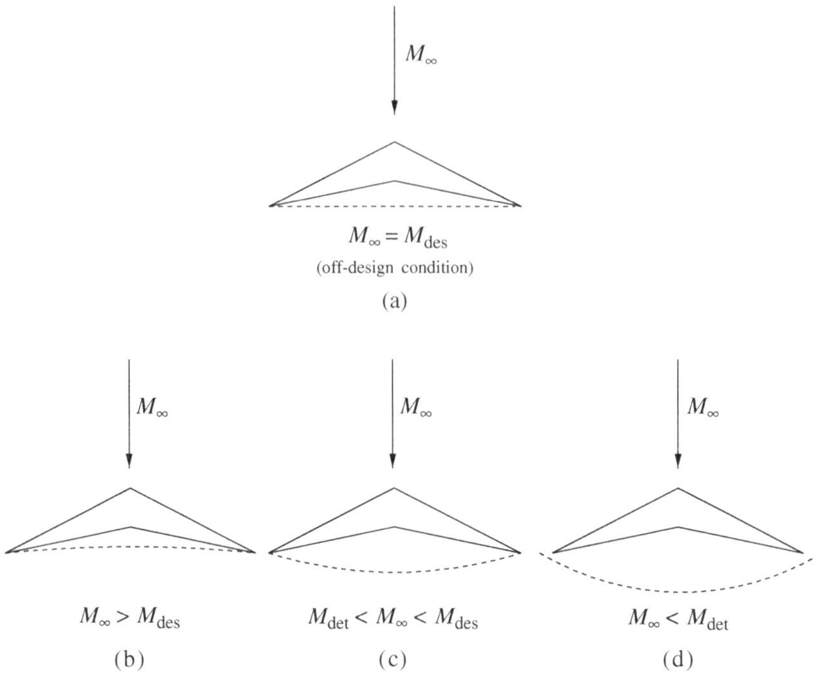

Figure 2.22 Shock shapes for caret-wings.

2.21 Approximate Theories for Caret-Wings

2.21.1 Two-Dimensional Wedge Theory for Caret-Wings

In this simplification, caret-wing is replaced by a wedge of the same base width $2s$, nose angle δ, freestream Mach number M_∞, and angle of incidence α, as shown in Fig. 2.23.

There are three cases of cruise for this wedge, as shown in Fig. 2.24, for the cases of angle of incidence: $\alpha > 0$ (Fig. 2.24(a)), $-\delta < \alpha < 0$ (Fig. 2.24(b)), and $\alpha < -\delta$ (Fig. 2.24(c)). These can be dealt with shock-expansion theory.

2.21.2 Wedge-Type Analysis

For the case of $0 < \alpha$, the pressure coefficient becomes

$$\frac{C_{pl}}{\delta^2} = \frac{\gamma+1}{2}\left[1 + \sqrt{1 + \frac{\left(\frac{4}{\gamma+1}\right)^2}{k^2\left(\frac{\alpha}{\delta}+1\right)^2}}\right]\left(\frac{\alpha}{\delta}+1\right)^2 \tag{2.27}$$

Figure 2.23 Slender wedge in a hypersonic flow.

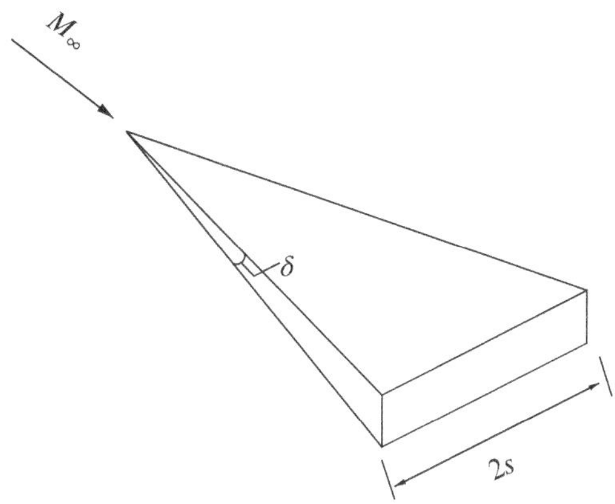

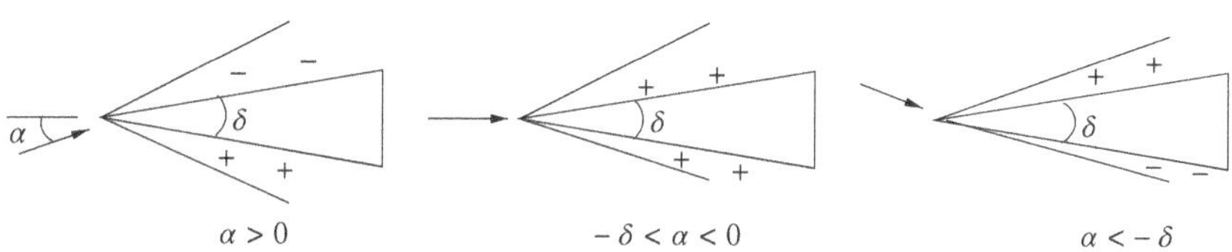

Figure 2.24 Slender wedge at different angles of incidence.

$$\frac{C_{pu}}{\delta^2} = \frac{2}{\gamma k^2}\left[\left(1 - \left(\frac{\gamma-1}{2}k\frac{\alpha}{\delta}\right)\right]^{2\gamma/(\gamma-1)} - 1\right] \tag{2.28}$$

where C_{pl} and C_{pu} are the pressure coefficients for the lower and upper surface of the wedge, respectively. Here y/s and **A** do not enter these equations because of the two-dimensional approximation.

For the case of $-\delta < \alpha < 0$, the C_{pl} is the same as that for $0 < \alpha$. The C_{pu} is given by

$$\frac{C_{pu}}{\delta^2} = \frac{\gamma+1}{2}\left[1 + \sqrt{1 + \frac{\left(\frac{4}{\gamma+1}\right)^2}{k^2\left(\frac{\alpha}{\delta}\right)^2}}\right]\left(\frac{\alpha}{\delta}\right)^2 \tag{2.29}$$

For $\alpha < -\delta$, the pressure coefficient C_{pl} is

$$\frac{C_{pl}}{\delta^2} = \frac{2}{\gamma k^2}\left[\left(1 - \frac{\gamma-1}{2}k\left(\frac{\alpha}{\delta}+1\right)\right)^{2\gamma/(\gamma-1)} - 1\right] \tag{2.30}$$

and C_{pu}/δ^2 is the same as that for $-\delta < \alpha < 0$.

2.21.3 Newtonian Approximation

By this approximation, for $\alpha > 0$, the shock is only on the lower surface, as illustrated in Fig. 2.25.

For the case of $-\sigma_{des} < \alpha < 0$, the shock on the wedge is on both sides, as shown in Fig. 2.26. The pressure coefficient for this case is

$$C_p = f(\alpha, \sigma_{des}, A)$$

For the case of $\alpha < -\sigma_{des}$, the shock is on the upper surface only and vacuum prevails on the lower surface, as illustrated in Fig. 2.27.

Thus, for $0 < \alpha$:

$$C_{pl} = 2\left(\alpha + \sigma_{des}\right)^2 \quad (2.30a)$$

$$C_{pu} = 0 \quad (2.30b)$$

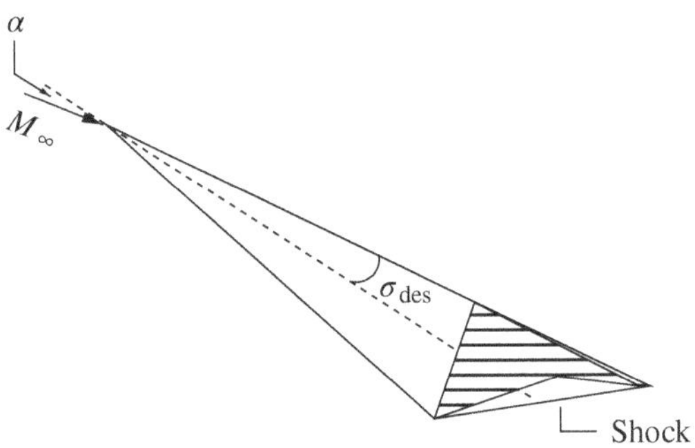

Figure 2.25 Slender wedge at positive angle of incidence.

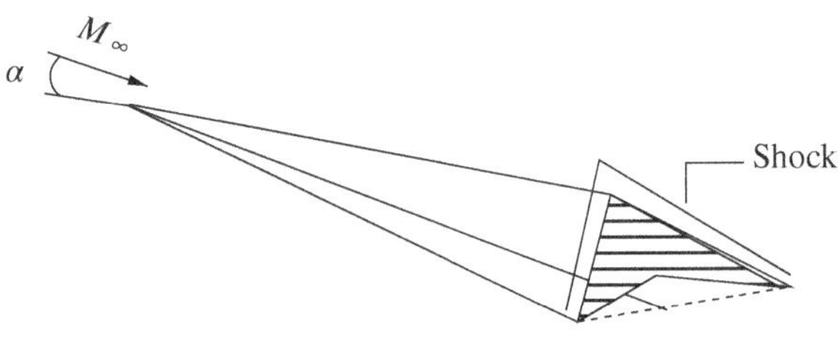

Figure 2.26 Wedge at angle of incidence between $\sigma_{des} < \alpha < 0$.

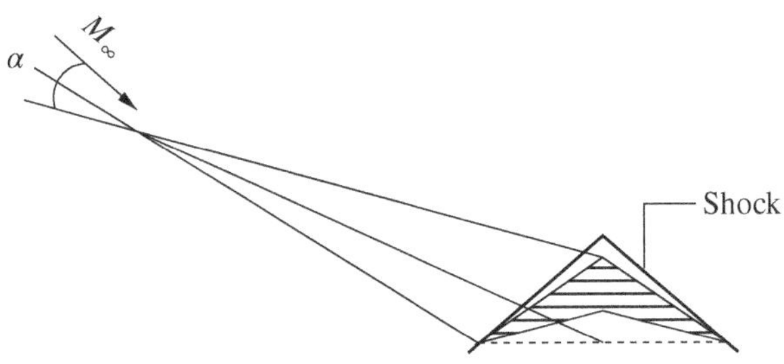

Figure 2.27 Wedge at an angle of incidence $\alpha < -\sigma_{des}$.

For $-\sigma_{\text{des}} < \alpha < 0$:

$$\frac{C_{pl}}{\delta^2} = 2\left(\frac{\alpha}{\delta} + \frac{\sigma_{\text{des}}}{\delta}\right)^2 \tag{2.31a}$$

$$\frac{C_{pu}}{\delta^2} = 2\frac{\left(\frac{\alpha}{\delta}\right)^2}{1 + 16\left(\frac{\sigma_{\text{des}}}{A}\right)^2} \tag{2.31b}$$

For $\alpha < -\sigma_{\text{des}}$:

$$C_{pl} = 0 \tag{2.32a}$$

$$\frac{C_{pu}}{\delta^2} = 2\frac{\left(\frac{\alpha}{\delta}\right)^2}{1 + 16\left(\frac{\sigma_{\text{des}}}{A}\right)^2} \tag{2.32}$$

None of these analysis give any idea about the span-wise pressure distribution, which may be necessary sometimes.

2.21.4 Extended Newtonian Formula

For the lower surface of a slender body, with $-\sigma_{\text{des}} < \alpha$, the pressure coefficient is given by [16–20],

$$\frac{C_{pl}}{\delta^2} = 2\left(\frac{\alpha}{\delta} + \frac{\sigma_{\text{des}}}{\delta}\right)^2 \left[1 + \epsilon\, p^* \left(\Omega,\, c,\, y/s\right)\right] \tag{2.33}$$

where

$$\epsilon = \frac{\gamma - 1}{\gamma + 1} + \frac{2}{(\gamma + 1)k^2\left(\frac{\alpha}{\delta} + \frac{\sigma_{\text{des}}}{\delta}\right)^2}$$

$$c = -\frac{\sigma_{\text{des}} - \delta}{\frac{A}{4}\sqrt{\epsilon}}$$

$$p^* = \frac{C_p - \left(C_p\right)_{\text{Newton}}}{\epsilon\left(C_p\right)_{\text{Newton}}}$$

$$\Omega = \frac{A/4}{\sqrt{\epsilon}\left(\alpha + \sigma_{\text{des}}\right)}$$

The term

$$\left[1 + \epsilon\, p^* \left(\Omega,\, c,\, y/s\right)\right]$$

in the C_p expression Eq. (2.33) derived for a general wing is valid for any body and not necessarily only for a slender body. The above equations are the specialized forms for slender wings.

Design Condition

$\Omega - c = -1$ corresponds to earlier formulae for design condition.

Detachment Condition

The pressure between the lower shock and the lower surface is calculated as $\Omega - c = 2$. For this case, the Mach number is given by

$$M_{\text{det}} = \frac{2}{\sigma_{\text{des}}}\left[\frac{\gamma+1}{2}\left(\frac{A}{4\sigma_{\text{des}}} + \frac{\sigma_{\text{des}} - \delta}{A/4}\right)^2 - 4(\gamma - 1)\right] \tag{2.34}$$

Equation (2.34) can be solved with computer, taking a number of points. Typical shock shapes and the pressure distribution are shown in Fig. 2.28.

Variation of Ω with c for the four cases to shock waves as illustrated in Fig. 2.28 is shown in Fig. 2.29.

2.21.5 Measurements Made at Braunschweing

Gun tunnel has the following advantages over a shock tunnel:

1) In gun tunnel, piston compresses the air. In shock tunnel, there is no piston, and the gas on the high-pressure side is following the shock. Therefore, this cold gas mixed very soon with the test gas.
2) The Reynolds numbers obtained in gun tunnel are higher than shock tunnel. This is because of the possibility of higher velocities and higher densities. Therefore, in a gun tunnel, higher temperatures and higher densities are possible to achieve. This enables the use of optical flow visualization such as Schlieren and interferometer easier.
3) All these are intermittent tunnels. Helium gas can be used as the test gas. The advantage of using helium is that its liquefaction temperature is very low. However, no real gas effects can be simulated. For example, helium being a noble gas, dissociation cannot be simulated. This is a disadvantage associated with the use of helium gas.
4) Heating is not uniform.
5) Rarefied gas study is possible because of the very low density in the test-section.
6) Measurements are difficult. Measurement of forces by indirect means like observing the path, observing the nose for temperature, and heat transfer are some of the usual practice. But measurements are very good from flow point of view. The following parametric variation, namely, the lift coefficient, C_L, drag coefficient, C_D, pitching moment coefficient, C_M as a function of incidence angle, δ_1, can be obtained comfortably.

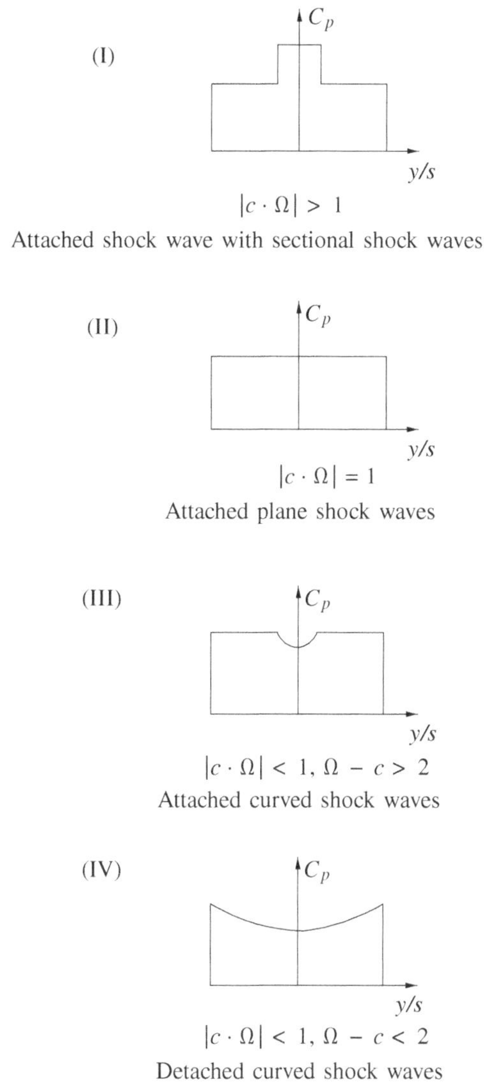

Figure 2.28 Shock shapes and the pressure distribution.

2.22 Supersonic Test Facilities

Some of the well-known supersonic test facilities are blowdown tunnel (without heating), blowdown tunnel (with heating), helium-tunnel, hot-shot tunnel, gun-tunnel, shock-tunnel, arc-heated tunnel, low-density tunnel, and ballistic-range.

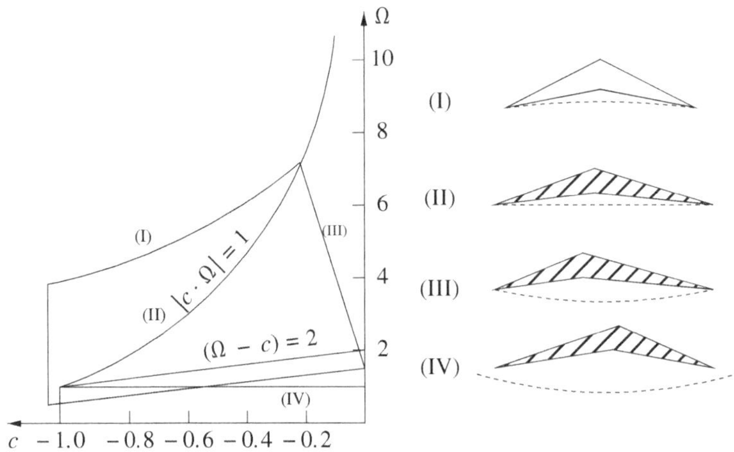

Figure 2.29 Ω variation with *c*.

2.22.1 Blowdown Tunnel (without heating)

The schematic sketch, operating parameters, and the application potential of a typical blowdown tunnel (without heating) are shown in Fig. 2.30.

2.22.2 Blowdown Tunnel (with heating)

The schematic sketch, operating parameters, and the application potential of a typical blowdown tunnel (with heating) are shown in Fig. 2.31.

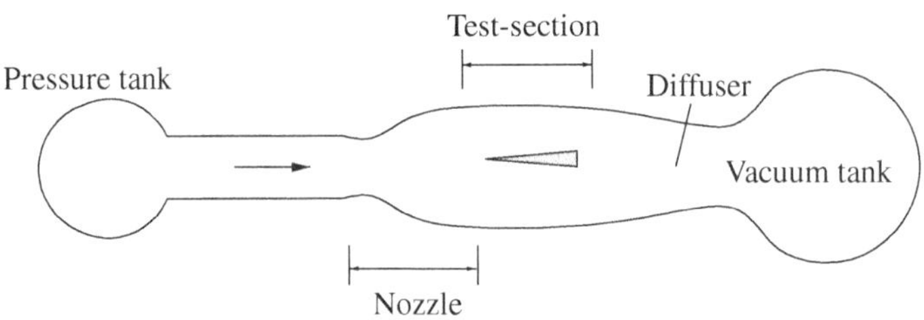

Mach number $M = 0.5$ to 3
Reynolds number $Re_{c\,1\,m} = 10^4 - 10^8$
Max. stagnation pressure $p_0 = 50$ atm
Max. stagnation temperature $T_0 = 300$ K
Running time: 60 sec.
Real gas effect: No
Use; flow studies, aerodynamic derivatives

Figure 2.30 Schematic sketch of a blowdown tunnel (without heating).

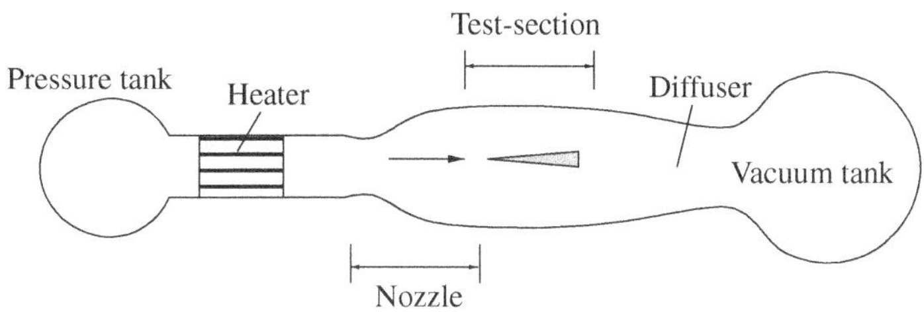

Mach number $M = 2$ to 10
Reynolds number $\text{Re}_{c\,1\,\text{m}} = 10^4\text{–}10^8$
Max. stagnation pressure $p_0 = 150$ atm
Max. stagnation temperature $T_0 = 1000$ K
Running time: 5–100 sec.
Real gas effect: No
Use; flow investigations, aerodynamic derivatives

Figure 2.31 Schematic sketch of a blowdown tunnel (with heating).

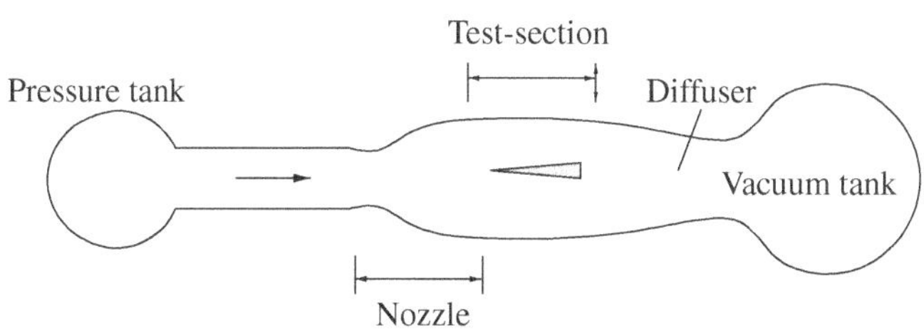

Mach number $M = 5$ to 25
Reynolds number $\text{Re}_{c\,1\,\text{m}} = 10^4\text{–}10^7$
Max. stagnation pressure $p_0 = 100$ atm
Max. stagnation temperature $T_0 = 300$ K
Running time: 5 to 100 sec.
Real gas effect: No
Use; flow studies (ideal gas)

Figure 2.32 Schematic sketch of helium tunnel.

2.22.3 Helium Tunnel

The schematic sketch, operating parameters, and the application potential of a typical helium tunnel are shown in Fig. 2.32.

The advantage of helium tunnel is that its liquefaction temperature is very low. The disadvantage is that no real gas effects, such as dissociation, can be simulated because it is a noble gas, therefore, does not dissociate.

2.22.4 Hot-Shot Tunnel

The schematic sketch, operating parameters, and the application potential of a typical hot-shot tunnel are shown in Fig. 2.33.

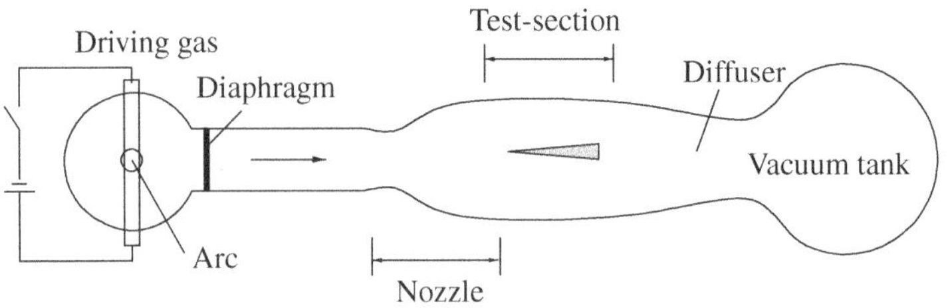

Mach number $M = 5$ to 25
Reynolds number $\mathrm{Re}_{c\,1\,\mathrm{m}} = 10^4 - 10^7$
Max. stagnation pressure $p_0 = 2000$ atm
Max. stagnation temperature $T_0 = 8000$ K
Running time: 0.020 to 0.100 sec.
Real gas effect: Yes
Use; flow studies, real gas effects, heat transfer

Figure 2.33 Schematic sketch of hot-shot tunnel.

The heating in the hot-shot tunnel is uniform. Therefore, the flow in the test-section is not homogeneous.

2.22.5 Gun-Tunnel

The schematic sketch, operating parameters, and the application potential of a typical gun tunnel are shown in Fig. 2.34.

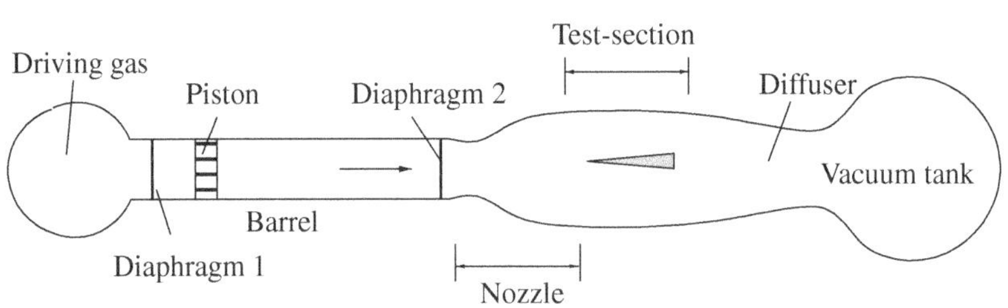

Mach number $M = 0.5$ to 13
Reynolds number $\mathrm{Re}_{c\,1\,\mathrm{m}} = 10^4 - 5 \times 10^7$
Max. stagnation pressure $p_0 = 200$ atm
Max. stagnation temperature $T_0 = 3000$ K
Running time: 0.100 sec.
Real gas effect: No
Use; flow studies, aerodynamic derivatives

Figure 2.34 Schematic sketch of gun tunnel.

Gun tunnel has the following advantages over a shock tunnel, without piston:

1) The piston in the gun tunnel compresses the air. In the shock tunnel, the gas on the high-pressure side follows the shock. Therefore, the cold gas gets mixed-up quickly with the test gas.
2) Reynolds numbers obtained are higher in the gun tunnel. This is because of the possibility of higher velocity and higher density. Therefore, in a gun tunnel, higher temperature and higher densities are possible in the test-section. Optical methods such as Schlieren are interferometer can be used for flow visualization.

2.22.6 Shock-Tunnel

The schematic sketch, the operating parameters, and the application potential of a typical shock tunnel are shown in Fig. 2.35.

2.22.7 Arc-Heated Tunnel

The schematic sketch, operating parameters, and the application potential of a typical arc-heated tunnel are shown in Fig. 2.36.

2.22.8 Low-Density Tunnel

The schematic sketch, operating parameters, and the application potential of a typical low-density tunnel are shown in Fig. 2.37.

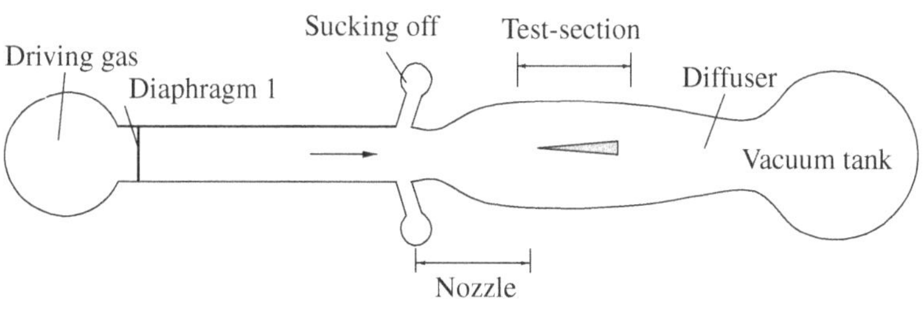

Mach number $M = 0.5$ to 13
Reynolds number $\mathrm{Re}_{c\,1\,\mathrm{m}} = 10^4 - 5 \times 10^7$
Max. stagnation pressure $p_0 = 200$ atm
Max. stagnation temperature $T_0 = 3000$ K
Running time: 0.100 sec.
Real gas effect: No
Use; flow studies, aerodynamic derivatives

Figure 2.35 Schematic sketch of shock tunnel.

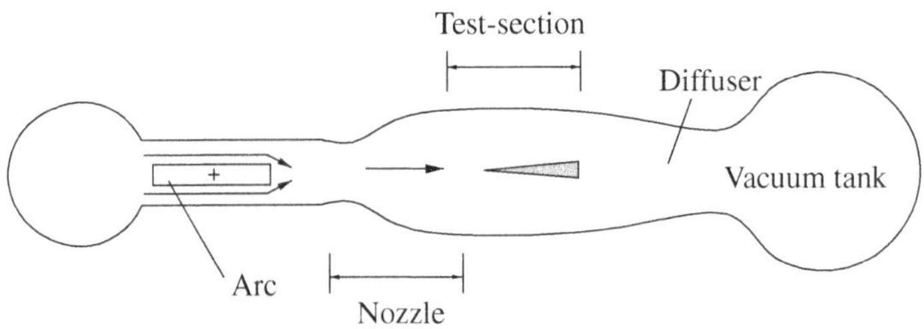

Mach number $M = 8$ to 22
Reynolds number $\mathrm{Re}_{c\,1\,\mathrm{m}} = 10^4 - 10^6$
Max. stagnation pressure $p_0 = 400$ atm
Max. stagnation temperature $T_0 = 6000$ K
Running time: 10–1000 sec.
Real gas effect: Yes
Use; real gas effects, ablation, heat transfer

Figure 2.36 Schematic sketch of arc-heated tunnel.

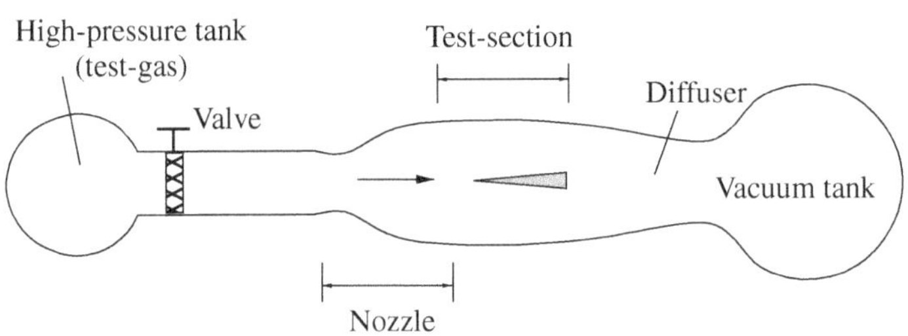

Mach number $M = 6$ to 12
Reynolds number $\mathrm{Re}_{c\,1\,\mathrm{m}} = 20 - 10^4$
Max. stagnation pressure $p_0 = 2$ atm
Max. stagnation temperature $T_0 = 4000$ K
Running time: continuous
Real gas effect: Yes
Use; Rarefied flow studies

Figure 2.37 Schematic sketch of a low-density tunnel.

All the test facilities listed above are intermittent, except the low-density tunnel. Rarefied gas study is possible because of the very low density in the test-section.

2.22.9 Ballistic Range

The schematic sketch, operating parameters, and the application potential of a typical ballistic range are shown in Fig. 2.38.

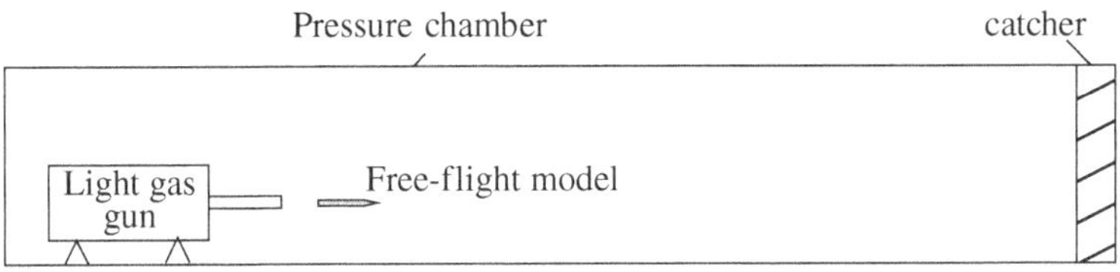

Figure 2.38 Schematic sketch of ballistic range.

Measurements are difficult in the ballistic range. Measurements are usually done by direct means like observing the path, observing the nose for temperature, and heat transfer. But the facility is very good from flow point of view.

2.22.10 Wedge Theory

This theory is very good for solving the problem of flow past a slender wedge in a freestream flow for the case of angle $\delta_1 > 0$, as shown in Fig. 2.39.

Angle δ_1 from the lower surface is given by

$$\delta_1 = \delta + \alpha$$

where α is measured from the upper surface of the wedge. Though the wedge theory is good for $\delta_1 > 0$, it is not good for $\delta_1 < 0$. This is because in wedge theory, the upper and lower surfaces are replaced by the plane surfaces. This approximation is reasonable for lower surface but bad for upper surface.

2.22.11 $(L/D)_{max}$ Variation with Freestream Mach Number M_∞

Variation of maximum aerodynamic efficiency, $(L/D)_{max}$, of the wedge with freestream Mach number is shown in Fig. 2.40.

It is seen that M_∞ has no effect on $(L/D)_{max}$. The decrease of $(L/D)_{max}$ is not due to M_∞ but is due to decrease of Reynolds number Re. That is, hypersonic flow boundary layer effects (viscous effects) become very important. Boundary layer effects become very important for hypersonic bodies in general, and for wave-riders in particular.

Figure 2.39 A wedge at an angle α to the freestream.

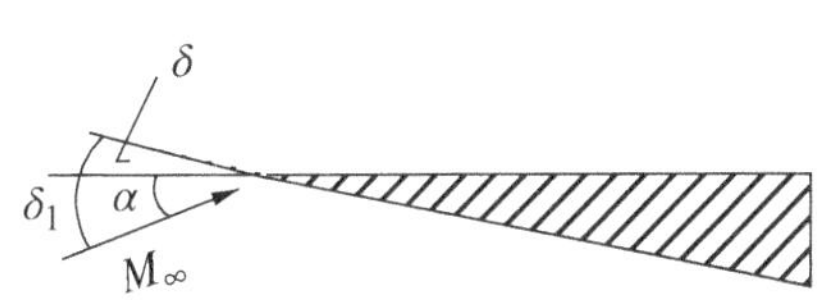

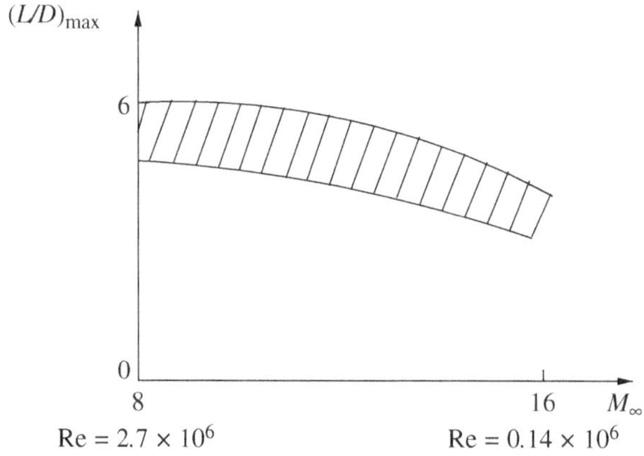

Figure 2.40 $(L/D)_{max}$ variation with M_∞.

2.23 Comparison of Theoretical and Experimental Results

Comparison of the experimental and theoretical results is shown in Fig. 2.41. It is seen that the theory agrees closely with the experiment. In this analysis, the viscous effects are taken into account by simple boundary layer calculations and taking displacement effects into consideration.

Variation of shock angle with angle δ_1 of a slender wedge presented in Fig. 2.41 shows a close agreement between the experimental and theoretical results.

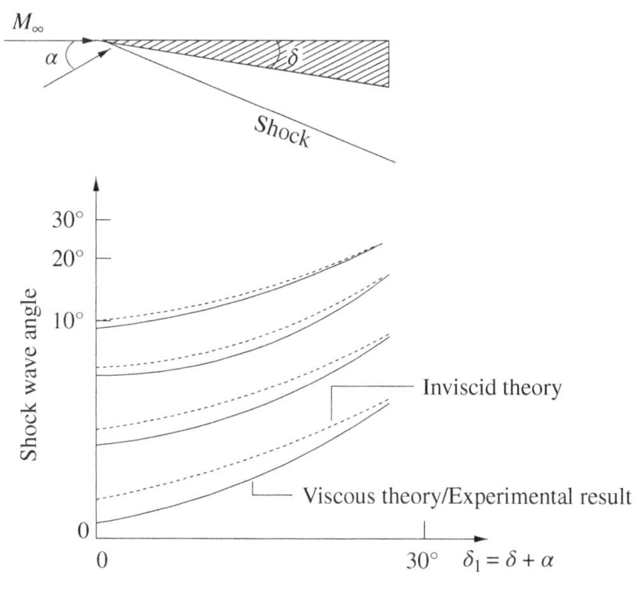

Figure 2.41 Shock angle variation with δ_1.

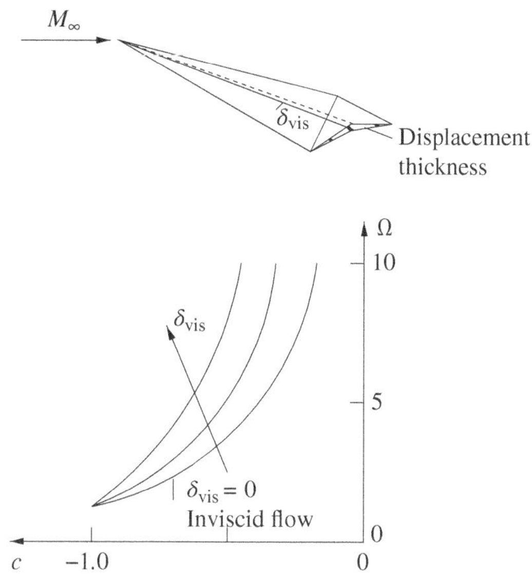

Figure 2.42 Displacement thickness variation with pitching moment Ω.

2.23.1 Pressure Distribution

Analysis of pressure coefficient, C_p, variation with y/s shows that even for design conditions, the shock is not in a plane but moved outward. This implies that there is a strong displacement effect of the boundary layer which pushes the shock outwards.

Displacement effects are more prominent at a low angle of attack, α. This is because of the relative increase in angle of attack due to boundary layer displacement which is more at low α. Therefore, the agreement is better at higher angle of attack, as shown in Fig. 2.42.

2.24 Stability Derivatives

Let us consider a slender delta wing shown in Fig. 2.43. Let

$$\lambda = c_0/c_i$$

$$F = \frac{1}{2}bc_i(1+\lambda)$$

$$C_L = \frac{L}{\frac{1}{2}\rho_\infty V_\infty^2 F}$$

$$\frac{x_{ac}}{c_i} = \frac{2}{3}(1-\lambda)$$

$$\mathbf{A} = \frac{b^2}{F} = \frac{2b}{c_i(1+\lambda)}$$

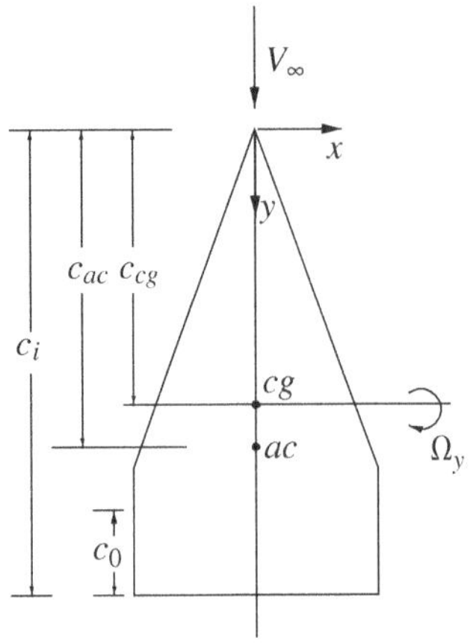

Figure 2.43 A slender delta-wing in an uniform flow.

$$C_\mu = \frac{2}{3}c_i\frac{1+\lambda+\lambda^2}{1+\lambda}$$

$$C_M = \frac{M}{\frac{1}{2}\rho_\infty V_\infty^2 F C_\mu}$$

$$\Omega_y = \frac{\omega_y C_\mu}{V_\infty}$$

In terms of the above parameters, the stability derivatives for delta wings (similar to that in Fig. 2.43) (the performance parameter derivatives) based on the theory of slender bodies with aspect ratio $\mathbf{A} < 1$ can be expressed as

$$C_{L_\alpha} = \frac{\partial C_L}{\partial \alpha} = \frac{\pi}{2}\mathbf{A} \tag{2.35}$$

$$C_{M_\alpha} = \frac{\partial C_M}{\partial \alpha} = -\frac{3}{4}\pi\mathbf{A}\frac{1+\lambda}{1+\lambda+\lambda^2}\left[\frac{2}{3}(1-\lambda) - \frac{x_{cg}}{c_i}\right] \tag{2.36}$$

$$C_{L_y} = \frac{\partial C_L}{\partial \Omega_y} = \frac{3}{4}\pi\mathbf{A}\frac{1+\lambda}{1+\lambda+\lambda^2}\left(1 - \frac{x_{cg}}{c_i}\right) \tag{2.37}$$

$$C_{M_y} = \frac{\partial C_M}{\partial \Omega_y} = -\frac{3}{8}\pi\mathbf{A}\frac{(1+\lambda)^2}{(1+\lambda+\lambda^2)^2}$$
$$\left[3\left(\frac{x_{cg}}{c_i}\right)^2 + (2\lambda-5)\frac{x_{cg}}{c_i} - \frac{9}{4} - \frac{3}{2}\lambda + \frac{3}{4}\lambda^2\right] \tag{2.38}$$

$$C_{L_{\hat{\alpha}}} = \frac{\partial C_L}{\partial\left(\frac{\hat{\alpha}\cdot c_\mu}{V_\infty}\right)} = \frac{1}{4}\mathbf{A}\frac{(1+\lambda)(1+2\lambda)}{1+\lambda+\lambda^2} \tag{2.39}$$

$$C_{M_{\hat{\alpha}}} = \frac{\partial C_M}{\partial\left(\frac{\hat{\alpha}\cdot c_\mu}{V_\infty}\right)}$$

or

$$C_{M_{\dot{\alpha}}} = \frac{3}{8}\pi\mathbf{A}\frac{(1+\alpha)^2}{(1+\alpha+\alpha^2)^2}\left[(1+2\lambda)\frac{x_{cg}}{c_i} - \left(\frac{3}{4} + \frac{3}{2}\lambda - \frac{3}{4}\lambda^2\right)\right] \tag{2.40}$$

$$C_y + C_{L_{\dot{\alpha}}} = \frac{3}{4}\pi\mathbf{A}\frac{1+\lambda}{1+\lambda+\lambda^2}\left(\frac{4+2\lambda}{3} - \frac{x_{cg}}{c_i}\right) \tag{2.41}$$

$$C_{M_y} + C_{M_{\dot{\alpha}}} = -\frac{9}{8}\pi\mathbf{A}\frac{(1+\alpha)^2}{(1+\alpha+\alpha^2)^2}\left(1 - \frac{x_{cg}}{c_i}\right)^2 \tag{2.42}$$

Maximum lift coefficient variation for slender wings is shown in Fig. 2.44, and the variation of maximum aerodynamic efficiency, (L/D), for slender wings is shown in Fig. 2.45.

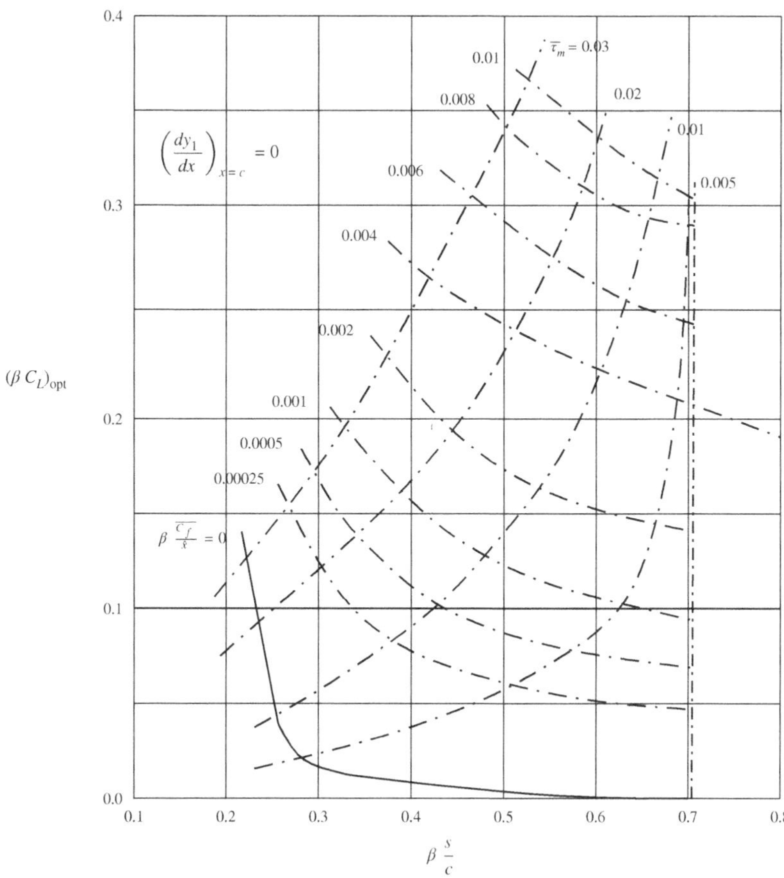

Figure 2.44 Maximum lift coefficient for slender wings.

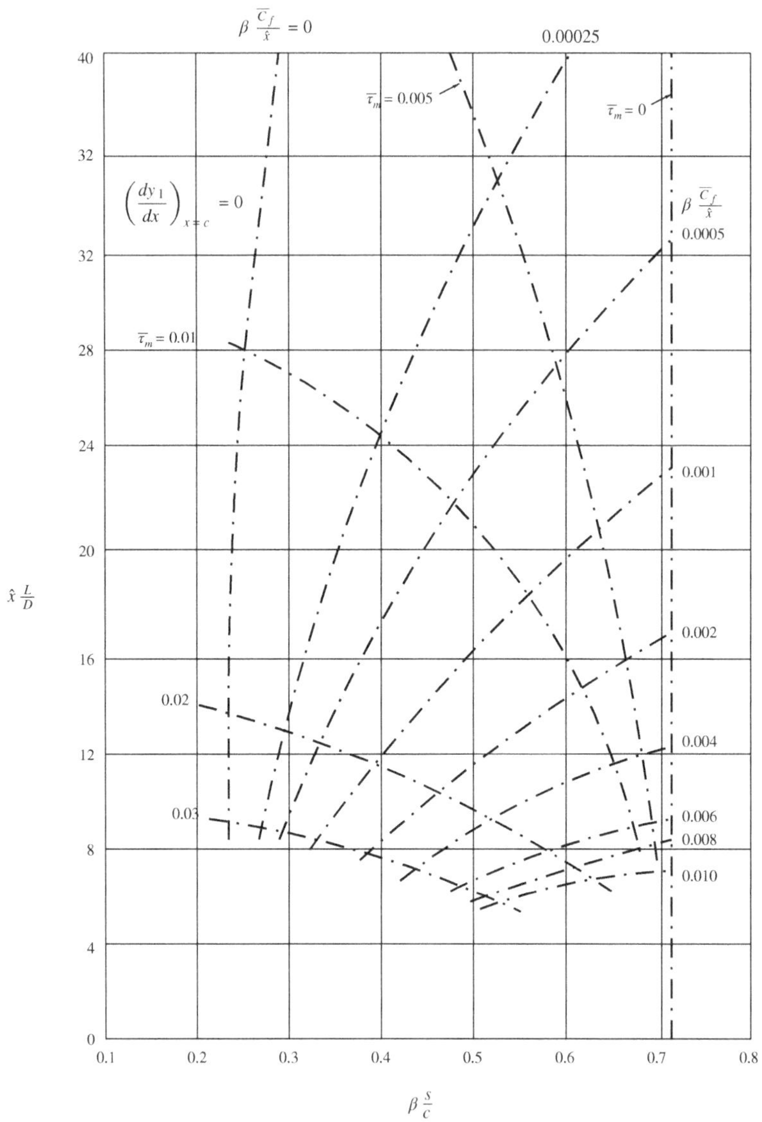

Figure 2.45 Maximum L/D-values for slender wings.

2.25 Summary

Hypersonic flow is defined as the flow with Mach number greater than 5, where the change in flow Mach number, M, is dictated by the change in the speed of sound, a. That is, in hypersonic flow regime, speed of sound, a, is more dominant than the flow speed, V, itself.

By definition,

$$\boxed{M_\infty \equiv \frac{V_\infty}{a_\infty} \gg 1}$$

The Mach number is greatly larger than unity ($M_\infty \gg 1$) is the basic assumption for all hypersonic flow theories. In flight applications, this results because the volume of the fluid particles is relatively large. The limiting case, where M_∞ approaches infinity because the freestream velocity approaches infinity while the freestream thermodynamic state remains invariant, produces extremely high temperatures in the shock layer.

The high temperatures associated with hypersonic flight are difficult to match in ground-test facilities, such as hypersonic wind tunnel, and shock tunnel. Therefore, in wind tunnel applications, hypersonic Mach numbers are achieved through relatively low speeds of sound. Thus, in hypersonic wind tunnels, the test-section Mach number, M_∞, approaches infinity because the speed of sound goes to *zero* while the freestream velocity is held constant.

Another assumption common to hypersonic flow is that the ratio of the freestream density to the density just behind a shock is extremely small, that is,

$$\boxed{\epsilon \equiv \frac{\rho_\infty}{\rho_2} \ll 1}$$

For slender body configurations, such as sharp-nosed cones and wedges, the strong shock assumption is

$$\boxed{M_\infty \sin\theta_b \gg 1}$$

where M_∞ is the freestream Mach number, and θ_b is the semi-angle of the nose. The concept is termed as the *Mach number-independence principle.*

For blunt bodies or slender bodies, at large angles of incidence, the pressure forces are much larger than the viscous forces, when the Reynolds number exceeds 10^5. Therefore, one would expect the Mach number independence principle to hold at these conditions.

There is a significant increase in the total drag coefficient for right circular cylinder due to the friction drag when the Knudsen number, Kn, defined as the ratio of mean-free path length, λ, to a characteristic dimension, d, in the flow field, is greater than 0.01.

The skin friction drag coefficient, in terms of Re_2, is given by

$$\boxed{C_{Df} = \frac{5.3}{\left(\mathrm{Re}_2\right)^{1.18}}}$$

For calculating the aerodynamic forces acting on a flying vehicle, it is necessary to know the atmospheric conditions prevailing around it.

The atmospheric properties that are of interest are the density of air, ρ, which is a scaling parameter to the aerodynamic coefficients via the dynamic pressure, $q = \frac{1}{2}\rho V^2$, and the speed of sound, a, which affects the Mach number, M, of the rocket, which in turn affects its aerodynamic properties. These may in turn be calculated from the air pressure, p, and temperature, T.

The characteristics of hypersonic flow can be computed using the fluid dynamic phenomena, when the configuration geometry, the altitude or orientation of the geometry to the freestream flow, and the altitude at which the vehicle flies or the test conditions of a wind tunnel simulation are known.

For hypersonic flow ($M_\infty > 5$), past slender bodies, the changes in flow properties are continuous; however, there are some perturbations. Therefore, it is not possible to linearize the equations, as in supersonic flow. This is because the fundamental equation governing most of the compressible flow regime

$$(1 - M_\infty^2)\,\phi_{xx} + \phi_{yy} + \phi_{zz} = 0$$

valid for subsonic, transonic, and supersonic flow under the framework of small perturbations with $u \ll V_\infty$, $v \ll V_\infty$, and $w \ll V_\infty$ is not valid for hypersonic flow even for slender bodies, since $u \approx V_\infty$ in the hypersonic flow regime.

Thus, the linear equation governing supersonic flow becomes nonlinear for hypersonic flows, because for hypersonic flow past slender bodies the assumption that the components of perturbation velocity; u, v, and w are small compared to freestream flow velocity, V_∞, and freestream speed of sound a_∞ is not valid. This is because for $M_\infty > 5$: (i) $V_\infty \approx 5\,a_\infty$ and (ii) $u \approx a_\infty/10$. Therefore, the second condition is satisfied even though the first condition is not violated.

In reality, for hypersonic flight, the temperature of the gas molecules that pass through the detached shock wave increases to very high levels, leading to the excitation of vibrational and chemical energy modes. This excitation lowers the specific heats ratio of the gas below the freestream value, if it is assumed that equilibrium exists and that dissociation is not driven to completion. A large amount of energies that would have gone into increasing the static temperature behind the bow shock wave for a perfect gas is used instead to excite the vibrational energy levels or to dissociate the gas molecules.

The compressibility factor is the ratio of the molecular weight of the undissociated air to the mean molecular weight at the conditions of interest.

For re-entry application, blunt-nosed body is preferable to avoid high stagnation temperature and to achieve efficient heat dissipation. But for sender bodies, one can go up to very high Mach numbers without taking into account real gas effects. There may be small regions near the leading edge of the body where the kinetic heating is large. Therefore, except at very high freestream Mach number, M_∞, there is no need to consider real gas effects.

For flying machines, such as aeroplane, when treated as slender bodies real gas effects are small. A re-entry satellite having freestream Mach number $M_\infty \sim 28$ will have to go through hypersonic–supersonic–subsonic regimes.

The flow turning angle–shock angle–Mach number (θ-β-M) relation for supersonic flow is

$$\frac{\tan(\beta - \theta)}{\tan \beta} = \frac{(\gamma - 1)M_1^2 \sin^2\beta + 2}{(\gamma + 1)M_1^2 \sin^2\beta}$$

Pressure coefficient for a hypersonic flow at any point in a flow is defined as

$$C_p = \frac{p - p_\infty}{\frac{1}{2}\rho_\infty V_\infty^2}$$

The dynamic pressure $q_\infty = \frac{1}{2}\rho_\infty V_\infty^2$ can expressed in terms of Mach number as

$$\begin{aligned} q_\infty &= \frac{1}{2}\frac{p_\infty}{a_\infty^2/\gamma} \\ &= \frac{\gamma}{2}p_\infty \frac{V_\infty^2}{a_\infty^2} \\ &= \frac{\gamma}{2}p_\infty M_\infty^2 \end{aligned}$$

Therefore, the pressure coefficient becomes

$$\begin{aligned} C_p &= \frac{p - p_\infty}{\frac{\gamma}{2}p_\infty M_\infty^2} \\ &= \frac{2}{\gamma M_\infty^2}\left(\frac{p}{p_\infty} - 1\right) \end{aligned}$$

The density ratio across the shock is

$$\frac{\rho_2}{\rho_1} = \frac{(\gamma + 1)M_\infty^2 \sin^2\beta}{(\gamma - 1)M_\infty^2 \sin^2\beta + 2}$$

In the limiting case, the freestream Mach number of a hypersonic flow tends to ∞, that is, $M_\infty \to \infty$, thus $M_\infty \theta \to \infty$. Therefore, in the limiting case the shock angle, β, pressure coefficient C_p/θ^2, and the density ratio ρ_2/ρ_∞ become

$$\beta = \frac{\gamma + 1}{2}\theta$$
$$\frac{C_p}{\theta^2} = \gamma + 1$$
$$\frac{\rho_2}{\rho_\infty} = \frac{\gamma + 1}{\gamma - 1}$$

For $\gamma = 1.4$,

$$\beta = 1.2\,\theta$$
$$\frac{C_p}{\theta^2} = 2.4$$
$$\frac{\rho_2}{\rho_\infty} = 6$$

Note that the density ratio maximum possible for the case of $M_\infty \to \infty$ is 6. This is the case for supersonic flow also.

For the limiting case of Newtonian flow, with the freestream Mach number tending to infinity ($M_\infty \to \infty$) and the specific heats ratio unity ($\gamma = 1$), the shock coincides with the surface of the body present in the flow.

To take into account, the effect of the freestream Mach number, M_∞, empirically, the Newtons formula depends on geometry only. At the leading edge of the body ($x = 0$), the pressure coefficient C_p is equal to that given by the oblique shock, and later on (i.e. downstream of the leading edge) the dependence is as per Newtonian formula,

$$C_p = F \cdot \frac{\theta_b(x)^2}{\theta_0^2}$$

where F is the pressure coefficient given by Eq. (2.11)

$$F = \frac{C_p}{\theta^2} = \frac{\gamma + 1}{4}\left[1 + \sqrt{\frac{\left(\frac{4}{\gamma+1}\right)^2}{M^2\theta^2}}\right]\theta_0^2$$

The similarity parameter is

$$\boxed{K = M_\infty \theta_0}$$

This K is known as *hypersonic similarity parameter*.

It is essential to note that the Newtonian theory is not good for some shapes. Experiments show that there is an appreciable difference in the pressure distribution predicted by the Newtonian model and the actual one. For both concave and convex body.

Busemann applied correction for the pressure distribution by considering the centrifugal force.

For more accurate analysis we need to use the method of characteristics. Consider the Prandtl–Meyer expansion of a flow of Mach number M_∞, turning through θ.

For this process, the Prandtl–Meyer function, ν, is the governing parameter,

$$\nu = \sqrt{\frac{\gamma+1}{\gamma-1}}\left[\tan^{-1}\sqrt{\frac{\gamma-1}{\gamma+1}(M^2-1)} - \tan^{-1}\sqrt{\frac{\gamma-1}{\gamma+1}(M_\infty^2-1)}\right]$$
$$-\tan^{-1}\sqrt{M_1^2-1} - \tan^{-1}\sqrt{M_\infty^2-1}$$

For similar flows, the optimization is done through the geometrical optimization. This kind of bodies is known as power-law bodies.

For two-dimensional bodies of shape given by $y_b \sim x^n$, the drag is given by

$$C_D = \frac{D}{q_\infty b\,\delta}$$

where $\delta = d/c$, d is the maximum thickness, b is the span, and c is the chord.

Theory of slender hypersonic bodies gives exact solution. It would capture the waves prevailing over the body, along with expansion wave reflection, etc.

The wave-rider is a typical three-dimensional slender body. A wave-rider is a hypersonic aircraft design that improves its supersonic lift-to-drag ratio by using the shock waves being generated by its own flight as a lifting surface (i.e. treating the waves coinciding with its surface), a phenomenon known as *compression lift*.

As the study of hypersonic flows improved, researchers were able to study wave-rider designs that used different shock wave shapes, the simplest being the conical shock generated by a cone.

The volume parameter, τ, is defined as

$$\tau = \frac{\mathbf{V}}{s^{3/2}} = \frac{2}{3}\frac{\delta}{\sqrt{A}}$$

Sometimes the approximate Newtonian formula, given below, is used.

The pressure coefficient is

$$C_{pl} = 2\,\delta_{\text{des}}^2$$

For the caret-wing, the pressure coefficient is given by

$$\frac{C_p}{\delta^2} = F\left(k, \frac{\alpha}{\delta}, A\,M_{\text{des}}, \tau\sqrt{M_{\text{des}}}, \gamma, \frac{y}{s}\right)$$

Shock detachment Mach number, M_{det}, is the freestream Mach number M_∞ at which the shock gets just detached from the leading edge.

Gun tunnel has the following advantages over a shock tunnel:

1) In gun tunnel, piston compresses the air. In shock tunnel, there is no piston and the gas on the high-pressure side is following the shock. Therefore, this cold gas mixed very soon with the test gas.
2) The Reynolds numbers obtained in a gun tunnel are higher than a shock tunnel. This is because of the possibility of higher velocities and higher densities. Therefore, in a gun tunnel, higher temperatures and higher densities are possible to achieve. This enables the use of optical flow visualization such as Schlieren and interferometer easier.
3) All these are intermittent tunnels. Helium gas can be used as the test gas. The advantage of using helium is that its liquefaction temperature is very low. However, no real gas effects can be simulated. For example, helium being a noble gas, dissociation cannot be simulated. This is a disadvantage associated with the use of helium gas.
4) Heating is not uniform.
5) Rarefied gas study is possible because of the very low density in the test-section.
6) Measurements are difficult. Measurement of forces by indirect means like observing the path, observing the nose for temperature and heat transfer are some of the usual practice. But measurements that are very good from flow point of view. The following parametric variation, namely, the lift coefficient, C_L, drag coefficient, C_D, pitching moment coefficient, C_M as a function of incidence angle, δ_1, can be obtained comfortably.

Some of the well-known supersonic test facilities are blowdown tunnel (without heating), blowdown tunnel (with heating), helium-tunnel, hot-shot tunnel, gun-tunnel, shock-tunnel, arc-heated tunnel, low-density tunnel, and ballistic range.

Wedge theory is good for solving the problem of flow past a slender wedge in a freestream flow for the case of angle $\delta_1 > 0$.

Exercise Problems

2.1 If the ratio of the downstream density to upstream density of the detached shock at the nose of a slender body is 4, find the specific heats ratio of the freestream flow.

[Ans. 1.67]

2.2 If the test model in a Mach 7 freestream has to obey the Mach number independence principle, with the constant as 4, determine the semi-angle of the model nose.

[Ans. 34.82°]

2.3 Assuming air as a perfect gas, determine the range of Mach number over which oxygen in air will dissociate at the sea level state.

[Ans. From 5.29 to 7.05]

2.4 Find the limiting Mach number at which the nitrogen in the air will begin to dissociate at the sea level.

[Ans. 13.22]

2.5 Find the freestream Mach number above which the atoms in the atmospheric air will begin to dissociate at the sea level, treating the air as a perfect gas.

[Ans. 26.45]

2.6 If the hypersonic similarity parameter for a slender body in freestream air of Mach number 7 is 0.8, determine (a) the wave angle at the shock formed at the nose of the body and (b) the nose angle.

[Ans. (a) 11.06°, (b) 6.55°]

2.7 Determine the shock distance from the base edge of a two-dimensional symmetrical wedge of nose angle 10° and height 2 m in a high hypersonic stream of the specific heats ratio as 1.2.

[Ans. 0.015 m]

2.8 Determine the percentage increase in the pressure coefficient on a slender body in a hypersonic limiting flow situation, when its nose angle is increased from 10° to 20°. Assume the specific heats ratio to be 1.

[Ans. 300%]

2.9 If the base pressure for a slender body of nose angle 10° is similar to that shown in Fig. 2.5 in a Newtonian limiting hypersonic flow of Mach number 33 at the sea level is 1 torr (abs), determine the base pressure coefficient, assuming the specific heats ratio as 1.

[Ans. 0.0276]

2.10 If the Newtonian drag experienced by a two-dimensional slender body represented by $y \sim x$ is 0.00125, determine the thickness-to-chord ratio of the body.

[Ans. 0.05]

2.11 If the pressure coefficient for a slender body of nose angle 2° in a Newtonian flow is 0.0025, determine the flow Mach number.

[Ans. 56]

2.12 A slender body of characteristic dimension 2 m flies at Mach 9 at 3 km altitude. Assuming the specific heats ratio as 1.4, determine the skin friction drag coefficient.

[Ans. 93.208×10^{-12}]

2.13 A caret-wing flies with aerodynamic efficiency 10. If the pressure coefficient is 0.05 and the specific heats ratio is 1.4, determine the design Mach number.

[Ans. 5.26]

2.14 A blunt-nosed slender body of diameter 1.2 m flies at Mach 3 at the sea level. If the wave drag experienced by the body is 430 kN, determine the wave drag coefficient.

[Ans. 0.597]

2.15 Determine the nose angle of the slender-coned body to fly at Mach 10 with the shock at its nose having the same strength as that at nose of a slender-coned body of nose angle 10° flying at Mach 4 in air. Assume the specific heats ratio as 1.4.

[Ans. 3°]

3

Application of Slender-body Theory

3.1 Introduction

Slender-body theory is used to calculate the effects on the lift and moment of mounting the wing of a wing–body combination above or below the body axis. The wing should have an unswept trailing edge and uncambered cross sections. Also, the wing must have a local span which increases in the downstream direction. For this geometry, the cross sections of the body are assumed to be circles of constant radius over the length of the wing.

The effects of the asymmetrical mounting can become substantial when the body diameter is more than half of the wing span. These effects will fall off as the body shrinks. This theory is particularly useful in the aerodynamic design considerations pertinent to the performance of slender-body vehicles operating at high velocities and high altitudes. In this analysis, the interaction between the viscous boundary layer and the inviscid flow field outside the boundary layer plays an important role. This interaction is found to have an important effect on the performance of the slender-body vehicles and vehicles meant for lifting entry from space.

The parametric group $(X_\epsilon/K_\epsilon^{2/3})$ that gives the relative importance of the bluntness of the body and viscous effects is found to be is approximately equal to [4]

$$\left(X_\epsilon/K_\epsilon^{2/3}\right) \approx 1.1\,\sqrt{(\lambda_\infty/t)}\,(x/t)^{1/6} \tag{3.1}$$

where λ_∞ is the mean free path of ambient air, t is the leading-edge thickness, and x is the streamwise distance. For values of $X_\epsilon/K_\epsilon^{2/3}$ less than 0.1, bluntness effects are dominant, and for $X_\epsilon/K_\epsilon^{2/3}$ greater than 1, the boundary layer displacement effects will dominate the inviscid flow field [4].

In the practical application of this theory, the following two important aspects to be addressed.

Hypersonic Slender Body Aerodynamics, First Edition. Ethirajan Rathakrishnan.

Companion Website: www.wiley.com/go/rathakrishnan/hypersonic

1) The limitation imposed by heat transfer on the trajectory of the vehicle.
2) The effect of body bluntness and viscous interaction on the lift and drag forces acting on the vehicle.

The lift and drag forces acting on a slender-body vehicle strongly depend on the relative effects of leading-edge bluntness and boundary layer displacement. At high altitudes, the Mach numbers and boundary layer thickness may be sufficiently large enough to cause streamline deflection produced by the boundary layer changes the thermodynamic state of the external flow. This coupling between the external inviscid flow and the viscous motion near the body surface is termed as *hypersonic viscous interaction.*

For solving the problem of the combined effect of the bluntness and viscous-interaction, several methods have been proposed. These studies revealed that the parametric group $(X_\epsilon/K_\epsilon^{2/3})$ that gives the relative importance of bluntness and viscous effects is approximately equal to Eq. (3.1). For $(X_\epsilon/K_\epsilon^{2/3}) \leq 1$, effects bluntness are dominant, whereas for $(X_\epsilon/K_\epsilon^{2/3}) \geq 1$ boundary layer displacement effect will dominate the inviscid flow field [4]. These expressions are given by [20]

$$X_\epsilon = \epsilon \sqrt{2} I_1 \frac{M_\infty^2 \sqrt{c_\infty}}{\mathrm{Re}_{x,\infty}}$$

$$K_\epsilon = \epsilon M_\infty^3 C_{D_w} \left(\frac{t}{x}\right)$$

where

$$\epsilon = \gamma - \frac{1}{\gamma} + 1$$

or

$$\epsilon = 2\xi \frac{d\beta}{d\xi}$$

where

$$\xi = \int_0^x \rho_w \mu_w u_e \, dx$$

For ballistic missiles,[1] I_1 is the displacement thickness integral, and β is the pressure gradient parameter given by

$$\beta = -1 \left(\frac{\gamma - 1}{\gamma}\right) \frac{\int_0^x p_x \, dx}{p_e^2} \left(\frac{dp_e}{dx}\right)$$

1 A rocket-propelled self-guided strategic weapons system that follows a ballistic trajectory (the trajectory of an object in free flight) to deliver a payload from its launch site to a predetermined target.

If the surface integration follows power law of the form $p \sim x^n$, then [4]

$$\beta = -\left(\frac{\gamma - 1}{\gamma}\right)\left(\frac{n}{1+1}\right)$$

subscript e in the expressions for ξ and β refers to value at the edge of the boundary layer and w refers to wall.

Example 3.1 A slender body of leading-edge thickness 5 mm and length 3 m cruises at sea level. Find whether the bluntness effects are negligible or dominant.

Solution

Given, $t = 5$ mm and $x = 3$ m. At the sea level, the mean free path of air is $\lambda_\infty = 0.1\ \mu$m.

By Eq. (3.1), we have that the bluntness effects are negligible if

$$1.1\sqrt{(\lambda_\infty/t)}\,(x/t)^{1/6} \ll 1$$

We have

$$\begin{aligned}
1.1\sqrt{(\lambda_\infty/t)}\,(x/t)^{1/6} &= 1.1 \times \sqrt{\frac{0.1 \times 10^{-6}}{5 \times 10^{-3}}} \times \left(\frac{3}{5 \times 10^{-3}}\right)^{1/6} \\
&= 1.1 \times \sqrt{0.00002} \times (600)^{1/6} \\
&= 1.1 \times \left(4.4721 \times 10^{-3}\right) \times 2.904 \\
&= 14.3 \times 10^{-3} \\
&= \boxed{0.0143}
\end{aligned}$$

Since the factor $1.1\sqrt{(\lambda_\infty/t)}\,(x/t)^{1/6} \ll 1$, the bluntness effects are negligible.

Example 3.2 Determine the flight altitude for a slender-body vehicle of leading-edge thickness of 10 mm and length of 5 m, at which the parameter $1.1\sqrt{(\lambda_\infty/t)}\,(x/t)^{1/6}$ becomes unity.

Solution

Given, $t = 10$ mm, $x = 5$ m, and $1.1\sqrt{(\lambda_\infty/t)}\,(x/t)^{1/6} = 1$.

Therefore,

$$\begin{aligned}
1.1\sqrt{(\lambda_\infty/t)} \times \left[5/(10 \times 10^{-3})\right]^{1/6} &= 1 \\
1.1\sqrt{(\lambda_\infty/t)} \times 2.817 &= 1 \\
\sqrt{(\lambda_\infty/t)} &= \frac{1}{1.1} \times \frac{1}{2.817} \\
&= 0.323 \\
\lambda_\infty/t &= 0.323^2 \\
&= 0.1043
\end{aligned}$$

$$\lambda_\infty = 0.1043 \times (10 \times 10^{-3})$$
$$= 1.043 \times 10^{-3}\ \text{m}$$

For $\lambda_\infty = 1.043 \times 10^{-3}$ m, from the atmospheric table, the altitude is about 100 km.

3.2 Leading-Edge Heat Transfer

The vehicle's ability to withstand high rates of heat transfer for a long period of time places severe limitations on the allowable trajectories. In most cases, the maximum heating rates can be calculated using blunt-body stagnation-point theories developed for the ballistic missiles[2] [4]. The validity of this approach is governed by the nature of the trajectory and the vehicle under consideration. For example, the maximum rate of heat transfer may be less than that predicted by stagnation-point theory by a factor of two or three.

At high altitudes, the number of collisions that occur near the body between incident molecules and the molecules that get reflected from the body surface becomes small. This is because the mean free path (i.e. the distance traveled by the molecules between successive collisions), λ, becomes large due to the low-density environment that prevails at high altitude. Therefore, the Knudsen number,[3] $\text{Kn} = \lambda/L$, where L is the characteristic dimension of the body (such as the diameter, d) becomes large. We know that, based on the range of Knudsen number, Kn, the flow is termed as continuum, transition, nearly free molecular, and free molecular.

That is,

for $\text{Kn} \leq 0.1$, the flow is termed as continuum.
for Kn in the range 0.1–1, the flow is termed as transition flow.
for Kn in the range 1–10, the flow is termed as nearly free molecular.
for $\text{Kn} > 10$, the flow becomes free molecular.

Once the flow ceases to be continuum, the problem has to be approached with a kinetic theory of gases. This kind of field is referred to as *low-density gas dynamics* or *rarefied gas dynamics*.[4] But as the vehicle descends through the atmosphere, a

---

2 A rocket-propelled self-guided strategic weapons system that follows a ballistic trajectory to deliver a payload from its launch site to a predetermined target.

3 A dimensionless number defined as the ratio of the molecular mean free path length to a representative physical length scale.

4 Rarefied gas flows are a classical discipline of fluid mechanics, which deals with the description of gases far from the state of equilibrium. Such flows are described with the Boltzmann equation. It is concerned with flows at such low density that the molecular mean free path is not negligible. Under these conditions, the gas no longer behaves as a continuum. Important modifications in aerodynamic and heat transfer characteristics occur which are ascribable to the basic molecular structure of the gas.

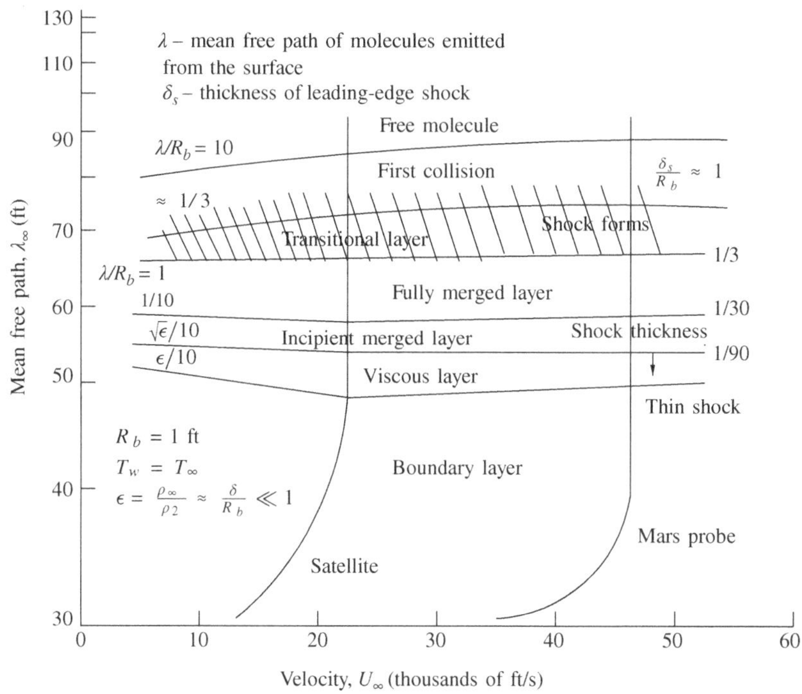

Figure 3.1 Flow regimes for a cooled sphere. Source: Hayes and Probstein [21]/with permission of ELSEVIER.

distinct shock wave forms ahead of the nose (i.e. leading edge). Due to the "cascading" effect of intermolecular collisions, this shock becomes stronger as the altitude decreases. This transition from free molecular to conventional boundary layer flow at high Reynolds number has been qualitatively shown in Fig. 3.1.

3.3 Stagnation-Point Heat Transfer

For lifting vehicles flying at sufficiently low altitudes where the characteristic leading-edge dimension is of the order of mean free path of freestream flow is the regime of at most interest. For the stagnation-point boundary layer, the heat transfer rate per unit area, q, is proportional to $R_b^{-1/2}$ [4]. In dimensionless form, the heat transfer can be expressed by Stanton number, St_o,[5] as

$$\mathrm{St}_o \equiv \frac{q}{\rho_\infty U_\infty (H_e - H_w)} = \frac{C_1}{\sqrt{\mathrm{Re}_o}} \tag{3.2}$$

5 A dimensionless number that measures the ratio of heat transferred into a fluid to the thermal capacity of fluid.

where

$$\mathrm{Re}_o = \frac{\rho_\infty U_\infty (2R_b)}{\mu_o}$$

R_b = nose radius

H_e = total enthalpy of the freestream

H_w = enthalpy of the gas at the body surface

μ_o = viscosity of the gas at freestream stagnation enthalpy H_e

C_1 is a function of the leading-edge shape, Mach number, degree of dissociation behind the shock wave at the nose, and body temperature, ρ_∞, and U_∞ are the freestream density and the velocity, respectively.

In a free molecular flow

$$\mathrm{St}_o = \alpha C_2 \tag{3.3}$$

where α is the energy-accommodation coefficient ($0 \leq \alpha \leq 1$) and C_2 is a function of Mach number, specific heats ratio, γ, and body shape.

For high Mach numbers, the relation between St_o and Re_o becomes independent of Mach number, in accordance with Mach number independence principle [11]. For this state, the variation of the difference between the Stanton number and the Reynolds number ($\mathrm{St}_o - \mathrm{Re}_o$), in principle, can be explicitly determined. For a circular cylinder normal to the flow direction at a high Mach number, the variation of ($\mathrm{St}_o - \mathrm{Re}_o$) with Reynolds number is shown in Fig. 3.2.

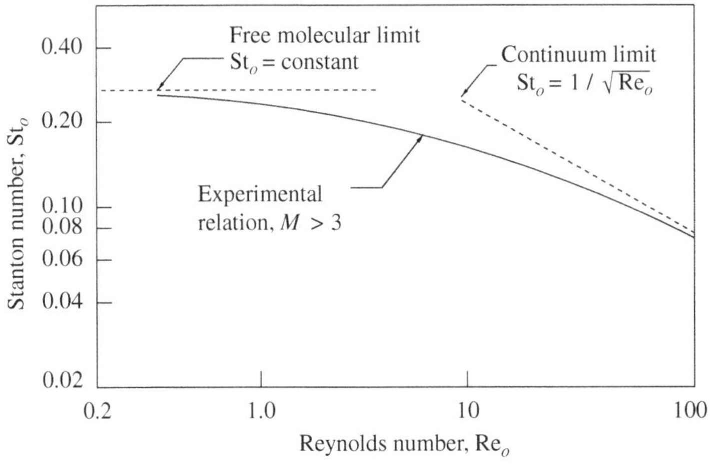

Figure 3.2 Heat transfer to a circular cylinder at low Reynolds number. Source: Rathakrishnan [12]/with permission of John Wiley & Sons.

From Fig. 3.2, it is seen that the stagnation point heat transfer rate, q, does not increase indefinitely as the leading-edge thickness is decreased, but rather asymptotically approaches a limiting value given by free molecular theory. For high Mach numbers, this limit is approximately given by

$$q \approx \alpha \left(\frac{1}{2}\rho_\infty U_\infty^2\right) U_\infty \sin\theta \left[1 - \frac{1}{\gamma}\left(\frac{\gamma+1}{\gamma-1}\right)\left(\frac{T_w}{T_\infty}\right)\frac{1}{M_\infty^2}\right] \tag{3.4}$$

where θ is the angle between the surface and the freestream direction.

Example 3.3 A slender-body vehicle flies at Mach 4.4 at 3000 m altitude. The thermal flux encountered by the vehicle at its nose is 500 W/m^2. If the Stanton number of the flow is 0.38, determine the temperature at the surface of the body, assuming the air as a perfect gas and the ratio of specific heats $\gamma = 1.4$.

Solution

Given, $M = 4.4$, $h = 3000$ m, $q = 500$ W/m^2, and $\text{St}_o = 0.38$.

From atmospheric table (Table A.2), at $h = 3000$ m, $T = 268.65$ K, and $\rho = 0.90929$ kg/m^3.

The speed of sound in the freestream is

$$\begin{aligned} a &= \sqrt{\gamma R T} \\ &= \sqrt{1.4 \times 287 \times 268.65} \\ &= 328.55\,\text{m/s} \end{aligned}$$

The freestream speed is

$$\begin{aligned} U &= M a \\ &= 4.4 \times 328.55 \\ &= 1445.62\,\text{m/s} \end{aligned}$$

The total temperature of the freestream is

$$\begin{aligned} T_0 &= T\left(1 + \frac{\gamma-1}{2}M^2\right) \\ &= T\left(1 + 0.2 \times 4.4^2\right) \\ &= 4.872\,T \\ &= 4.872 \times 268.65 \\ &= 1308.86\,\text{K} \end{aligned}$$

The total enthalpy of the freestream flow is

$$\begin{aligned} H_0 &= c_p T_0 \\ &= 1004.5 \times 1308.86 \\ &= 1.315\,\text{MJ/kg} \end{aligned}$$

By Eq. (3.2),

$$\mathrm{St}_o \equiv \frac{q}{\rho_\infty U_\infty (H_e - H_w)}$$

Therefore,

$$0.38 = \frac{500}{0.90929 \times 1445.62 \times (H_0 - H_w)}$$

$$H_0 - H_w = \frac{500}{0.38 \times 0.90929 \times 1445.62}$$

$$\begin{aligned}
&= 1.001\\
H_w &= H_0 - 1.001\\
&= 1315000 - 1.001\\
&= 1314999\\
c_p T_w &= 1314999\\
T_w &= \frac{13149999}{c_p}\\
&= \frac{1314999}{1004.5}\\
&= \boxed{1309.11\ \mathrm{K}}
\end{aligned}$$

3.4 Heat Transfer Limitations for Slender-body Vehicles

Ballistic missile is a typical slender-body vehicle meant for operation under extreme heat transfer condition for a few seconds. But a typical vehicle which is capable of flying for many minutes in high-temperature environment is known as high-altitude lifting vehicle. The most promising one among the mechanisms developed for withstanding this kind of environment involves the use of ablating surfaces to reduce the magnitude of the heating rate, thermal radiation to dissipate the absorbed heat energy, or a combination of these two [4].

For estimating heat transfer under a wide range of re-entry conditions, the correlation developed by Gazley is found to be very useful [4]. Gazley's formula for axisymmetric bodies is [20]

$$\overline{q_F} \approx \frac{3}{2}\sqrt{\frac{M_\infty}{\mathrm{Re}_{\infty,d}}}\,\rho_\infty U_\infty^3 \tag{3.5}$$

where $\overline{q_F}$ is the average rate at which heat is transferred to the nose per unit frontal area of the body. Including both two-dimensional and axisymmetric bodies, this relation may be rearranged as[6] [4]

$$\frac{\overline{q_F}\sqrt{R_b}}{\sqrt{p(o)}}\left[\frac{f(\Lambda)}{\sqrt{2}}\right]^{j-1} \approx 0.05H_e \frac{\text{Btu}}{\text{ft}^{3/2}\ \text{atm sec}} \tag{3.6}$$

where

R_b = nose radius, ft

$p(o)$ = pressure at the stagnation point

$H_e = \frac{U_\infty^2}{2}$ = stagnation enthalpy of freestream, Btu/lb

$f(\Lambda)$ = leading-edge sweep parameter $\approx (\cos\Lambda)^{3/2}$

Λ = leading-edge angle

For a two-dimensional leading edge, the superscript j in Eq. (3.6) is unity (i.e. $j = 0$), and for axisymmetric bodies, $j = 1$ [4].

The heat transfer distribution near the leading edge of slender bodies with hemispherical or hemicylindrical leading edge is approximately represented by [4]

$$\frac{q(\theta)\sqrt{R_b}}{\sqrt{p(o)}}\left[\frac{f(\Lambda)}{\sqrt{2}}\right]^{j-1} \approx C_1 H_e \cos\theta \tag{3.7}$$

For Gazleys's analysis, the constant $C_1 \approx 0.05$ and from the correlation of Scala [20] $C_1 \approx 0.036$, $q(\theta)$ is the heat-transfer rate per unit surface area at an angle θ from the axis of symmetry.

Slender vehicles with high lift/drag ratio require a small nose radius R_b. In a continuum flow, the heat-transfer rate is found to be proportional to $R_b^{-1/2}$, suggesting that leading-edge heat transfer may be an important limiting design consideration [4].

Let us consider a material capable of withstanding high-temperature and high-emissivity radiation, such as graphite.[7]

An estimate of the equilibrium temperature may be obtained by equating the convective heat transfer to the heat radiated from the surface.

6 $\text{Btu}/(\text{ft}^2\ \text{s}) = 11\,356.5385\ \text{W/m}^2$.

7 Graphite is a mineral composed of stacked sheets of carbon atoms with a hexagonal crystal structure. It is the most stable form of pure carbon under standard conditions. Graphite is very soft, has a low specific gravity, is relatively nonreactive, and has high electrical and thermal conductivities.

For graphite, the convective heat transfer at the stagnation point, from Eq. (3.7) with $C_1 = 0.036$ is [4]

$$\frac{q(o)\sqrt{R_b}}{\sqrt{p(o)}}\left[\frac{f(\Lambda)}{\sqrt{2}}\right]^{j-1} = 0.36\,H_e \tag{3.8}$$

while the surface radiative heat flux is

$$q_r = 0.476\,\epsilon\left(\frac{T}{1000}\right)^4 \text{Btu/(ft}^2\text{ s)} \tag{3.9}$$

Here T is the surface temperature in Rankine and ϵ is the surface emissivity. Equating q_r and $q(o)$, we get [4]

$$T = 0.480\times 10^3\left(\frac{\sqrt{p(o)}\,H_e}{\epsilon\,\sqrt{R_b}}\right)^{1/4} \tag{3.10}$$

The parameters in Eq. (3.10) are the same as those in Eq. (3.6).

Missions requiring long times of exposure to the temperature conditions of radiation equilibrium are so many. Although materials such as graphite are capable of withstanding these surface temperature, severe surface mass loss and heat penetration to the supporting structure may occur. Scala [20] proposed the following correlation for diffusion-controlled mass loss from a graphite surface [4]

$$\frac{\dot{m}_w\sqrt{R_b}}{\sqrt{p_e}}\left[\frac{f(\Lambda)}{\sqrt{2}}\right]^{j-1} = 6.35\times 10^{-3}\ \text{lb/(ft}^{3/2}\text{ s atm}^{1/2}) \tag{3.11}$$

Pressure p_e in Eq. (3.11) is the local surface pressure in the atmosphere. Assuming a density of 135 lb/ft^3 ($\approx$16.0185 kg/m^3), for the material, the maximum surface-erosion rate occurring at the stagnation point can be expressed as

$$\frac{\dot{r}_w\sqrt{R_b}}{\sqrt{p_e}}\left[\frac{f(\Lambda)}{\sqrt{2}}\right]^{j-1} = 4.66\times 10^{-5}\ \text{lb/(ft}^{3/2}\text{ s atm}^{1/2}) \tag{3.12}$$

It is seen that the mass loss is independent of surface temperature in the diffusion-controlled regime. Thus, the mass loss is considerably less than that which would occur with low-temperature ablators such as quartz. Some typical stagnation-point erosion rates for an unswept two-dimensional leading edge are listed in Table 3.1 [4].

The thermal conductivity of several proposed high-temperature radiators to the surface is sufficient to make sustained high surface temperatures impractical [4]. However, new pyrolytic materials such as pyrolytic graphite[8] have highly anisotropic properties, and the conductivity normal to the surface is closer to the efficient insulator.

---

8 Pyrolytic graphite sheets are ultra-thin and lightweight graphite films with a very high thermal conductivity.

Table 3.1 Graphite erosion rates for an unswept two-dimensional leading edge. Adapted from [4].

			$V_w\sqrt{R_b}$ ($ft^{3/2}/s \times 10^{-5}$)	
			U_∞ (ft/s)	
Altitude (ft)	**17 000**	**20 000**	**23 000**	**26 000**
160	1.9	2.3	2.6	2.9
180	1.34	1.6	1.8	2.1
200	0.79	1.1	1.3	1.5
220	0.66	0.77	0.89	1.0
240	0.43	0.51	0.58	0.65
260	0.27	0.31	0.35	0.41
280	0.15	0.17	0.20	0.22

3.5 Sublimation

Advantages of both radiative and mass-transfer cooling are offered by several high-temperature materials. However, in contrast to the case of diffusion-limited mass transfer where the exothermic combustion reaction takes place at the surface at a rate determined by the diffusion of oxygen to the surface, the sublimation rate is an exponentially increasing function of temperature, and the combustion reaction occurs within the boundary layer [4].

In addition to the heat absorbed during the sublimation process, further reductions in heat transfer are achieved by virtue of the gaseous effusion away from the surface. This "blowing" substantially reduces the net heat transfer to the surface [4].

3.6 Aerodynamic Effects

Aerotropic thermal conductivity, such as that possessed by pyrolytic graphite, allows the shape of a sharp leading edge and a constant increase in the lift/drag ratio in addition to providing effective thermal insulation for the supporting structure. The severe thermal stresses caused by large temperature gradients normal to the surface must be carefully studied to protect the vehicle.

For ablating bodies, the aerodynamic profile will change with time. If the leading-edge thickness, t, is initially of the order of a few millimeter would tend to become a centimeter, this thickness may change by a factor or two or more during sustained high-altitude flight. For a blunt-nosed vehicle, say $t = 2$ and $R_b = 0.5$ m, the percentage change in t due to ablation would be small for materials like graphite. Since the lift, drag, and heat transfer are strong functions of the leading-edge thickness, it is important to carefully study the rate of ablation.

At high surface temperatures, T_w, the thermal boundary layer thickness increases. At high Mach number and altitude, the boundary layer may be sufficiently thick to cause an appropriate distortion of the external flow field. This coupling is referred to as *hypersonic viscous interaction*.[9] It is found that the ratio of wall temperature to stagnation temperature, T_w/T_o, has a strong influence on the surface pressure and heat transfer.

3.7 Nose Bluntness

In a continuum regime of flow, the increase of leading-edge thickness results in reduced heat-transfer rate. But this results in considerable increase of drag, resulting in the decrease of aerodynamic efficiency. Degradation of aerodynamic efficiency with an increase of bluntness for some typical slender bodies can be inferred from the variation of surface pressure variation and $C_D/C_{p_{\max}}$ along the length of the body shown in Figs. 3.3–3.5.

Example 3.4 A blunt-nosed slender body flies at a sea level. If the freestream Mach number is 4.5, find the pressure coefficient at the stagnation point.

Solution

Given, $M_\infty = 4.5$. For the blunt-nosed body, there will a detached shock at the nose. Let subscripts 1 and 2 refer to states ahead of and behind the shock.

The pressure coefficient at the stagnation point can be expressed as

$$C_{p_0} = \frac{p_0 - p_\infty}{q_\infty}$$

where p_0 is the pressure at the stagnation point, p_∞ is the freestream static pressure, and q_∞ is the freestream dynamic pressure.

The dynamic pressure q_∞ can be expressed as

$$\begin{aligned} q_\infty &= \frac{1}{2}\rho_\infty V_\infty^2 \\ &= \frac{1}{2}\left(\frac{p_\infty}{RT_\infty}\right) V_\infty^2 \\ &= \frac{1}{2}\left(\frac{\gamma p_\infty}{\gamma RT_\infty}\right) V_\infty^2 \end{aligned}$$

9 Coupling between the external inviscid flow and the viscous motion near the body surface is termed as hypersonic viscous interaction.

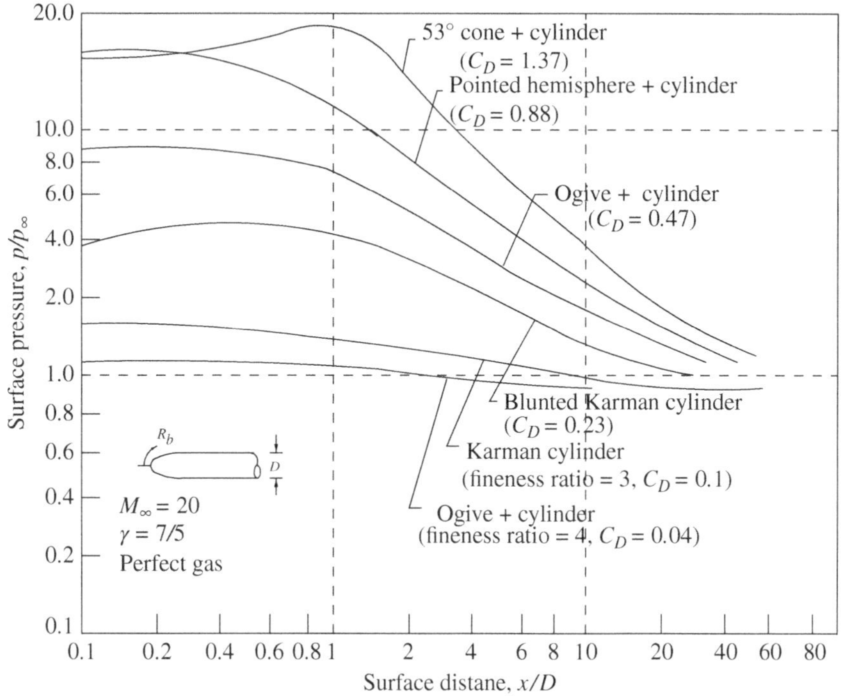

Figure 3.3 Surface pressure, p/p_∞, variation along the surface. Source: Adapted from Squire [20].

$$= \frac{1}{2}\left(\frac{\gamma p_\infty}{a_\infty^2}\right) V_\infty^2$$

$$= \frac{1}{2}\gamma p_\infty M_\infty^2$$

Therefore,

$$C_{p_0} = \frac{p_\infty}{\frac{1}{2}\gamma p_\infty M_\infty^2}\left(\frac{p_0}{p_\infty} - 1\right)$$

But the stagnation pressure here is p_{02}, the total pressure behind the shock at the nose. Therefore,

$$C_{p_0} = \frac{2}{\gamma M_\infty^2}\left(\frac{p_{02}}{p_\infty} - 1\right)$$

Now, for $M_\infty = 4.5$, from normal shock table, $p_{02}/p_{01} = 0.09$, and from isentropic table, $p_\infty/p_{01} = 0.0035$. Therefore,

$$\frac{p_{02}}{p_\infty} = \frac{p_{02}}{p_{01}}\frac{p_{01}}{p_\infty}$$

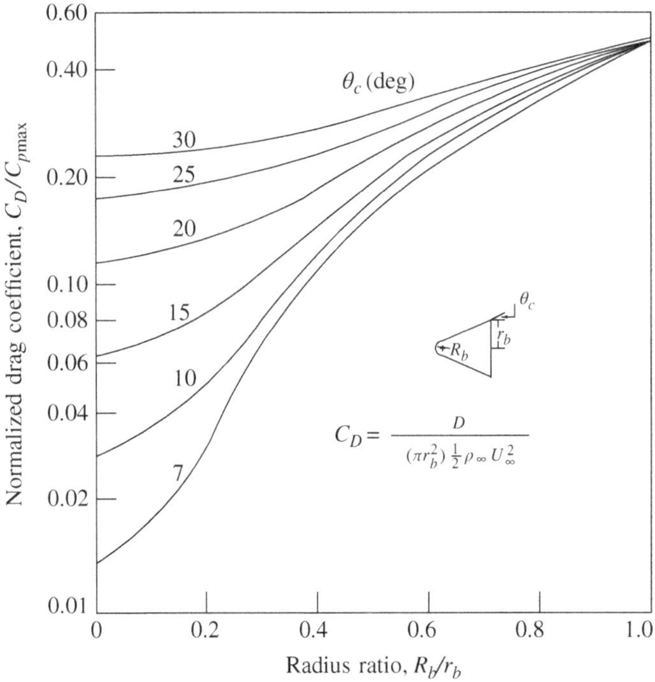

Figure 3.4 Drag coefficient of spherical-nosed cone modified with Newtonian pressure coefficient. Source: Adapted from Dewey Jr. [4].

$$= 0.09 \times \frac{1}{0.0035}$$
$$= 25.7$$

The pressure coefficient is

$$C_{p_{02}} = \frac{2}{1.4 \times 4.5^2} \times (25.7 - 1)$$
$$= 0.0705 \times 24.7$$
$$= 1.74$$

3.8 Blast-wave Theory

Widely used methods for the prediction of pressure distributions on blunt-nosed slender bodies is the blast-wave theory and the related power law. The *equivalence principle* proposed by Hayes and Probstein [21] is the basis for these results. Steady

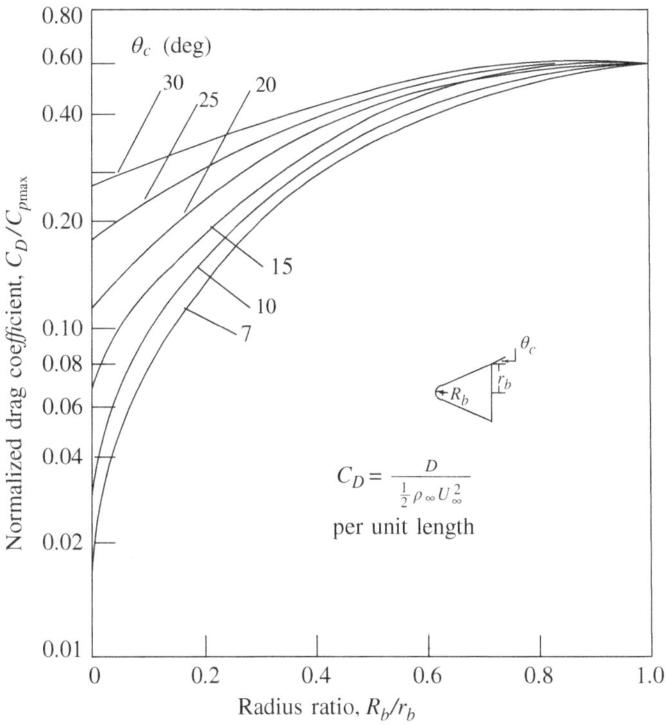

Figure 3.5 Drag coefficient of cylindrically blunted wedge using modified Newtonian pressure distribution.

flow over a slender body at hypersonic speeds is equivalent to an unsteady flow in one space dimension is the essence of equivalence principle. This principle replaces the time coordinate in the unsteady problem with x/U_∞ in the corresponding steady hypersonic flow. This requires $|(u - U_\infty)/U_\infty| \ll 1$ everywhere in the inviscid field. For an oblique shock with small shock angle, β, in a freestream flow of large Mach number, M, and $M\beta \geq O(1)$,

$$\frac{u - U_\infty}{U_\infty} \approx -\frac{2\beta^2}{\gamma + 1}\left(1 - \frac{1}{M^2\beta^2}\right) \ll 1 \tag{3.13}$$

Near the leading edge of a blunt body, where the shock angle β is of order unity, the equivalence principle fails locally.

Example 3.5 A slender body of nose angle 10° flies at Mach 8. Assuming the specific heats ratio as 1.4, check whether equivalence principle is valid for this flight.

Solution

Given, $M = 8$, $\theta = 10°$, and $\gamma = 1.4$. For $M = 8$, and $\theta = 10°$, from oblique shock chart, the shock angle $\beta = 16°$.

Therefore, by Eq. (3.13),

$$\begin{aligned}\frac{u - U_\infty}{U_\infty} &\approx -\frac{2\beta^2}{\gamma+1}\left(1 - \frac{1}{M^2\beta^2}\right)\\ &= -\frac{2 \times (16/57.3)^2}{1.4+1} \times \left(1 - \frac{1}{8^2 \times (16/57.3)^2}\right)\\ &= -0.065 \times (1 - 0.2)\\ &= \boxed{-0.052}\end{aligned}$$

that is, $(u - U_\infty)/U_\infty \ll 1$, therefore, the equivalence principle is valid.

3.9 Thin Shock Layers

Solutions of the blast-wave solutions fall under the general classifications of small perturbation theory. The blast-wave results in desirable property that the solution is always self-similar in the normalized transverse distance. For more general problems which do not correspond to blast-wave requirement, the small perturbation equations have to be solved numerically for any affine family of bodies. However, it should be emphasized that the difficulty in applying small perturbation theory to the combined bluntness and viscous-induced external field is an important issue that needs to be addressed. This is because the effective body shape is not known prior to the calculation of the boundary layer, and the boundary-layer thickness depends on the resulting effective body shape. This coupling may be circumvented in the following ways [4]:

1) A flow field associated with the growth of the boundary layer over the original body can be determined by assuming that the boundary layer produces only a small perturbation on the flow field.
2) An additional approximation to the small disturbance equations assumes the ratio of specific heats, γ, is close to unity.

A large part of the flow field between the body and the shock wave is encompassed by the entropy layer, within this thin layer, the pressure rapidly decreases from its value directly behind the shock to the value which pertains across the entropy layer.

By imposing the pressure field, the boundary conditions along the body surface, and the external entropy layer, the growth of the boundary layer is determined. For plane flow, in the limit, the specific heats ratio $\gamma \to 1$, the complete region may be calculated from the local shape of the outer edge of the entropy layer. Also, the velocity $u(x) \approx U_\infty$ is everywhere in the entropy layer in this approximation.

The lift and drag acting on the body are strongly influenced by the viscous interaction. An increase of surface pressure associated with viscous interaction is significantly larger than the increase in skin friction. The regime where the effects of leading-edge bluntness and viscous interaction are of comparable magnitude is the zone of practical interest for slender bodies.

3.10 Summary

Slender-body theory is used to calculate the effects on the lift and moment of mounting the wing of a wing–body combination above or below the body axis. The wing should have an unswept trailing edge and uncambered cross sections. Also, the wing must have a local span which increases in the downstream direction.

The effects of the asymmetrical mounting can become substantial when the body diameter is more than half of the wing span.

The parameter $\left(X_\epsilon/K_\epsilon^{2/3}\right)$ is approximately equal to

$$\left(X_\epsilon/K_\epsilon^{2/3}\right) \approx 1.1\sqrt{(\lambda_\infty/t)}\,(x/t)^{1/6}$$

defines the relative importance of bluntness and viscous effects.

In the practical application of this theory, the following two important aspects to be addressed.

1) The limitation imposed by heat transfer on the trajectory of the vehicle.
2) The lift and drag of the vehicle are strongly influenced of bluntness and viscous interaction.

The lift and drag of a slender vehicle strongly depend on the relative effects of leading-edge bluntness and boundary layer displacement.

The coupling between the external inviscid flow and the viscous motion near the body surface is termed as *hypersonic viscous interaction*.

Several methods have been proposed for solving the problem of the combined effect of the bluntness and viscous interaction. These studies revealed that the parameter defining the relative importance of bluntness and viscous effects is $\left(X_\epsilon/K_\epsilon^{2/3}\right)$, and for flight conditions this parameter is approximately equal to Eq. (3.1).

Ballistic missile is a rocket-propelled self-guided strategic weapons system that follows a ballistic trajectory (the trajectory of an object in free flight) to deliver a payload from its launch site to a predetermined target.

The vehicle's ability to withstand high rates of heat transfer for a long period of time places severe limitations on the allowable trajectories. In most cases, the maximum heating rates can be calculated using blunt body stagnation point theories developed for the ballistic missiles.

At high altitudes, the number of collisions that occur near the body between the incident molecules and the molecules that get reflected from the body surface becomes small.

For Kn ≤ 0.1, the flow is termed as continuum.
For Kn in the range 0.1–1, the flow is termed as transition flow.
For Kn in the range 1–10, the flow is termed as nearly free molecular.
For Kn $>$ 10, the flow becomes free molecular.

Once the flow ceases to be continuum, the problem has to be approached with kinetic theory of gases. This kind of field is referred to as *low-density gas dynamics* or *rarefied gas dynamics*.

For lifting vehicles, flying at sufficiently low altitudes where the characteristic leading-edge dimension is of the order of mean free path of freestream flow is the regime of at most interest. For the stagnation-point boundary layer, the heat transfer rate per unit area, q, is proportional to $R_b^{-1/2}$. In dimensionless form, the heat transfer can be expressed as Stanton number, as

$$\mathrm{St}_o \equiv \frac{q}{\rho_\infty U_\infty (H_e - H_w)} = \frac{C_1}{\sqrt{\mathrm{Re}_o}}$$

For high Mach numbers, the relation between St_o and Re_o becomes independent of Mach number, in accordance with Mach number independence principle.

The ballistic missile is a typical slender-body vehicle meant for operation under extreme heat transfer condition for a few seconds. But a typical vehicle which is capable of flying for many minutes in high-temperature environment is known as high-altitude lifting vehicle. The most promising one among the mechanisms developed for withstanding this kind of environment involves the use of ablating surfaces to reduce the magnitude of the heating rate, thermal radiation to dissipate the absorbed heat energy, or a combination of these two.

The heat transfer distribution near the leading edge of hemispherical or hemicylindrical leading edge is given by

$$\frac{q(\theta)\sqrt{R_b}}{\sqrt{p(o)}} \left[\frac{f(\Lambda)}{\sqrt{2}}\right]^{j-1} \approx C_1 H_e \cos\theta$$

For Gazleys's analysis, the constant $C_1 \approx 0.05$ and from the correlation of Scala [20] $C_1 \approx 0.036$, $q(\theta)$ is the heat-transfer rate per unit surface area at an angle θ from the axis of symmetry.

Slender vehicles with a high lift/drag ratio require a small nose radius R_b. In a continuum flow, the heat-transfer rate is found to be proportional to $R_b^{-1/2}$, suggesting that leading-edge heat transfer may be an important limiting design consideration.

There are many missions requiring long times of exposure to the temperature conditions of radiation equilibrium. Although materials such as graphite are capable of withstanding these surface temperature, severe surface mass loss and heat penetration to the supporting structure may occur.

In the diffusion-controlled regime, the mass loss is independent of surface temperature. Thus, the mass loss is considerably less than that which would occur with low-temperature ablators such as quartz.

The thermal conductivity of several proposed high-temperature radiators to the surface is sufficient to make sustained high surface temperatures impractical. However, new pyrolytic materials such as pyrolytic graphite have highly anisotropic properties, and the conductivity normal to the surface is closer to efficient insulator.

Several high-temperature materials offer the advantages of both radiative and mass-transfer cooling. However, in contrast to the case of diffusion-limited mass transfer where the exothermic combustion reaction takes place at the surface at a rate determined by the diffusion of oxygen to the surface, the sublimation rate is an exponentially increasing function of temperature, and the combustion reaction occurs within the boundary layer [4].

In addition to the heat absorbed during the sublimation process, further reductions in heat transfer are achieved by virtue of the gaseous effusion away from the surface. This "blowing" substantially reduces the net heat transfer to the surface.

Aerotropic thermal conductivity, such as that possessed by pyrolytic graphite, allows the shape of a sharp leading edge and a constant increase in the lift/drag ratio in addition to providing effective thermal insulation for the supporting structure.

For ablating bodies, the aerodynamic profile will change with time. If the leading-edge thickness, t, is initially of the order of a few millimeter would tend to become a centimeter, this thickness may change by a factor or two or more during sustained high-altitude flight.

At high surface temperatures, T_w, the thermal boundary layer thickness increases. At high Mach number and altitude, the boundary layer may be sufficiently thick to cause an appropriate distortion of the external flow field. This coupling is referred to as *hypersonic viscous interaction*. The ratio of wall temperature to stagnation temperature, T_w/T_0, has a strong influence on the surface pressure and heat transfer.

In a continuum regime of flow, the increase of leading-edge thickness results in a reduced heat-transfer rate. But this results in considerable increase of drag, resulting in the decrease of aerodynamic efficiency.

Widely used methods for the production of pressure distributions on blunt-nosed slender bodies is the blast-wave theory and the related power law. The *equivalence principle* proposed by Hayes and Probstein [21] is the basis for these results. Steady flow over a slender body at hypersonic speeds is equivalent to an unsteady flow in one space dimension is the essence of equivalence principle.

The blast-wave solutions are only one family of solutions falling under the general classifications of small perturbation theory. The blast-wave results in a desirable property that the solution is always self-similar in the normalized transverse distance. For more general problems which do not correspond to blast-wave requirement, the small perturbation equations have to be solved numerically for any affine family of bodies.

The viscous interaction is found to have considerable effect on the lift and drag. The relative increase of surface pressure, due to viscous interaction, is much larger than the relative increase in skin friction. A regime of significant practical interest for slender bodies exists in which the effects of leading-edge bluntness and viscous interaction are of comparable magnitude.

Exercise Problems

3.1 If a slender body of nose thickness 10 mm and length 2.6 m flies at an altitude where the mean free path is 4×10^{-7} m, find whether the bluntness effects are significant or insignificant.

[Ans. The bluntness effects are significant]

3.2 If a slender body of nose thickness 20 mm and streamwise distance 4 m flies in air, with $1.1\sqrt{(\lambda_\infty/t)}\,(x/t)^{1/6} = 2$, determine the mean free path.

[Ans. 0.0113 m]

3.3 If the perturbation velocity of the flow field generated by a blast wave is 10% of the flow speed, determine the flow Mach number assuming the specific heat ratio as 1.2.

[Ans. 10]

3.4 If the perturbation velocity associated with a blast wave is 30% of the wave speed, determine the Mach number of the blast wave, assuming $\gamma = 1.2$.

[Ans. 2.09]

3.5 If the average heat transfer rate to the nose per unit area of a slender body of characteristic dimension is 1 m, flying with freestream temperature is 200 K, and density 0.8 kg/m^3 is 6 MW/m^2, determine its flight Mach number. Assume $\gamma = 1.2$.

[Ans. 10.325]

3.6 A slender body of characteristic dimension 3 m flies at Mach 12, with freestream density and temperature 0.9 kg/m^3 and 260 K, respectively. Assuming the specific heats ratio as 1.4, find the average heat transfer rate to the nose per unit area.

[Ans. 10.82 MW/m^2]

3.7 A slender body of nose angle 20° flies at Mach 6 and angle of attack of 3°, with freestream density and temperature 1.18 kg/m^3 and 220 K, respectively. Assuming the specific heats ratio as 1.4, find the average heat transfer rate to the nose per unit area.

[Ans. 1.44 MW]

3.8 Experimental result of a slender body flying in free molecular state, with Mach 6, gives the Reynolds number as 1.2. (a) Find the proportionality constant which depends on the leading-edge shape and (b) the proportionality constant between the Stanton number and Reynolds number, if the energy accommodation coefficient is 0.8.

[Ans. (a) 0.219 and (b) 0.25]

3.9 The freestream density and temperature during reentry of an axisymmetric vehicle are 0.72 kg/m^3 and 180 K, respectively. If the characteristic dimension of the axisymmetric vehicle is 0.02 m and the flight Mach number is 9, determine the heat transfer per unit nose area, assuming the gas constant as 287 m^2/(s^2 K) and the ratio of specific heats as 1.1.

[Ans. 20 MW/m^2]

4

Experimental Approach

4.1 Introduction

One of the typical slender bodies we see commonly is wings. Wings are only one of the types of slender body that can be tested in wind tunnel. Like wings that play a crucial role in aeronautical applications (as well as in some non-aeronautical applications), other slender bodies, such as aircraft fuselages and engine nacelles, also play equally crucial role. These bodies are not designed to produce lift although they can do. Their main aerodynamic contribution is drag. Some of the well-known slender bodies for non-aeronautical use are street luminaries, wind turbine nacelles, and anemometers.

4.2 Drag of Slender Bodies

There are no simple methods available for estimating the aerodynamic forces acting on slender bodies. Slender bodies are usually designed for operation at minimum drag and zero lift. Therefore, the main consideration in the design of slender bodies is the estimation of the drag of such bodies. Nevertheless, they can generate significant amounts of lift in off-design conditions.

The best approach for estimating the drag of slender bodies is to carry out wind tunnel experiments. There are some simple methods that can be used in the design stage. However, these can prove to be highly inaccurate. Such methods are usually applied in the conceptual design stage for aircraft.

Drag components consist of skin friction drag, profile drag, induced drag, and interference drag. For slender bodies, the main drag sources are the skin friction drag and profile drag. Slender bodies being streamlined shapes, the profile drag is

Hypersonic Slender Body Aerodynamics, First Edition. Ethirajan Rathakrishnan.

Companion Website: www.wiley.com/go/rathakrishnan/hypersonic

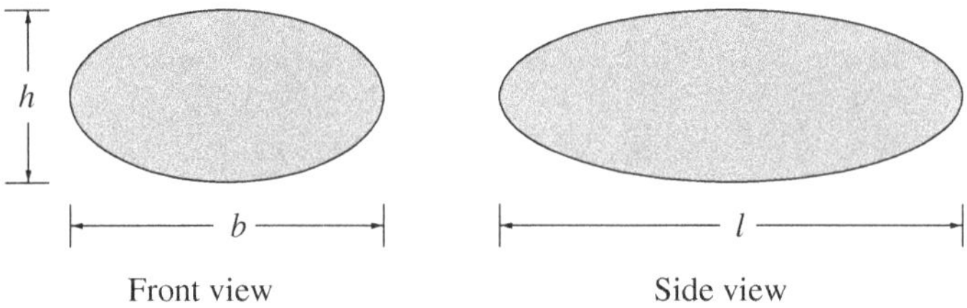

Figure 4.1 A typical slender body.

lower than the skin friction drag. Also, drag due to interference and compressibility effects may be ignored.

Let us consider a typical slender body of length l, width b, and height h as shown in Fig. 4.1.

The drag of fuselages, tail booms, and other slender bodies can be estimated using the flat plate analogy. The skin friction drag of the body is approximated by that of a flat plate with the same wetted area. The flow regime (laminar or turbulent) over the flat plate must be a representative of the real flow. A shape correction is applied in order to model the profile drag. This shape correction compares the shape of the body to that of a cylindrical "cigar."

4.2.1 Flat Plate Analogy

By flat plate analogy, the profile drag of the body is given by

$$C_D S = C_f(1 + \phi)S_{\text{wet}} \tag{4.1}$$

where C_D is the drag coefficient, S is the reference area, S_{wet} is the wetted area of the body, and ϕ is the shape correction factor. The drag coefficient can be expressed as

$$C_D = \frac{D}{\frac{1}{2}\rho_\infty V_\infty^2 S}$$

where D is the drag, ρ_∞ and V_∞, respectively, are the density and velocity of the freestream flow, and S is the reference area.

4.2.1.1 Flat Plate Solution for Skin Friction

For a laminar flow, the skin friction coefficient of a flat plate is given by [22].

$$C_f = \frac{1.328}{\sqrt{(\text{Re})_x}} \tag{4.2a}$$

where Re is the Reynolds number based on the length of the slender body. This law of friction deduced by Blasius is valid only for the laminar flow, i.e. for $(\text{Re})_x <$ (5×10^5) to 10^6. In the region of turbulent motion with $(\text{Re})_x > 10^6$, the drag becomes considerably greater than that given by Eq. (4.2a).

For a turbulent flow, the skin friction coefficient is,

$$C_f = \frac{0.072}{[(\mathrm{Re})_x]^{1/5}} \tag{4.2b}$$

This expression is valid for Reynolds number up to 10^7, but the experimental results indicate that a better approximation is given by

$$C_f = \frac{0.074}{[(\mathrm{Re})_x]^{1/5}} \tag{4.2c}$$

Prandtl has suggested subtracting the length of the laminar layer, resulting in the expression,

$$C_f = \frac{0.074}{[(\mathrm{Re})_x]^{1/5}} - \frac{1700}{(\mathrm{Re})_x} \tag{4.2d}$$

which can be applied for $(\mathrm{Re})_x$ from (5×10^5) to 10^7.

To extend the Reynolds number range further, Schlichting and Gersten [23] employed the logarithmic velocity for a pipe under turbulent flow condition resulting in the semi-empirical relation,

$$C_f = \frac{0.455}{[\log_{10}(\mathrm{Re})_x]^{2.58}} \tag{4.2e}$$

Equation (4.2*e*) can be used for Reynolds number greater than 10^7. For $\mathrm{Re} < 10^7$, Eq. (4.2*e*) gives the value for C_f that are very close to those given by Eq. (4.2*d*), and, consequently, engineers commonly use Eq. (4.2*e*) over the entire range of Reynolds numbers above 500 000.

For streamlined bodies, the separated zone being small, the major portion of the drag is from skin friction.

For streamlined bodies, the separated zone being small, the major portion of the drag is because of skin friction. Bodies are classified as streamlined and bluff, based on which is dominant among the drag components. A body for which the skin friction drag is the major portion of the total drag is termed as *streamlined body*. A body for which the pressure (form) drag is the major portion is termed as *bluff body*. Turbulent boundary layer results in more skin friction than laminar one. Examine the skin friction coefficient, C_f, variation with Reynolds number, for a flat plate kept at zero angle of attack in a uniform stream, plotted in Fig. 4.2. The characteristic length for Reynolds number is the plate length x, from its leading edge. It can be seen from Fig. 4.2 that the C_f is more for turbulent flow than laminar flow. The skin friction coefficient is defined as:

$$C_f = \frac{\text{Frictional force}}{\frac{1}{2}\rho_\infty V_\infty^2 S}$$

where V_∞ and ρ_∞ are the freestream velocity and density, respectively, and S is the wetted surface area of the flat plate.

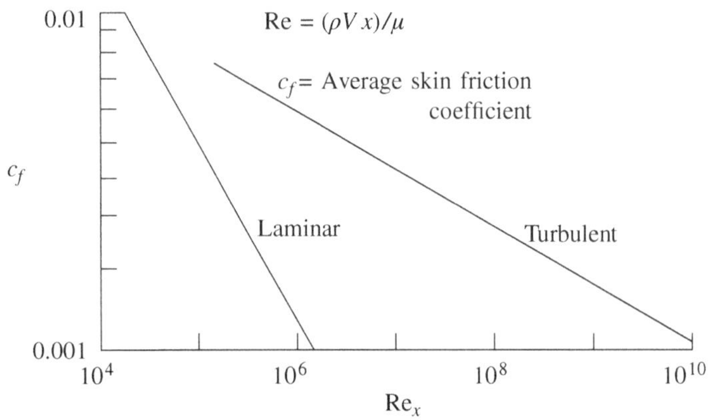

Figure 4.2 Variation of C_f with Reynolds number.

For bluff bodies, the pressure drag is substantially greater than the skin friction drag, and, for streamlined bodies, the condition is the reverse. In the case of streamlined bodies, such as aerofoil, the designer aims at keeping the skin friction drag as low as possible. Maintaining laminar boundary layer conditions all along the surface is the most suitable arrangement to keep the skin friction low. Though such aerofoils, known as laminar aerofoils, have been designed, they have many limitations. Even a small surface roughness or disturbance can make the flow turbulent and spoil the purpose of maintaining the laminar flow over the entire aerofoil. In addition, for laminar aerofoils, there is a tendency for the flow to separate even at small angles of attack, which severely restricts the use of such aerofoils.

If the flow is not fully turbulent, it will start off as laminar and undergo turbulent transition over the plate. Then, the skin friction coefficient will depend on the location of the transition point. For high Reynolds numbers, it is simpler to assume that the flow is fully turbulent. For low Reynolds numbers, both the laminar and the turbulent drag can be calculated. Then, the actual drag will lie between these two values. The skin friction coefficient will also be affected by compressibility effects.

4.2.1.2 Slenderness Ratio

Consider the slender body of cross-sectional area A_c, length $l = l_N + l_c + l_A$, with faired forebody and afterbody, as shown in Fig. 4.3.

The slenderness ratio of the body is defined as

$$\lambda_{\text{eff}} = \frac{l}{\sqrt{(4/\pi)A_c}} \tag{4.3a}$$

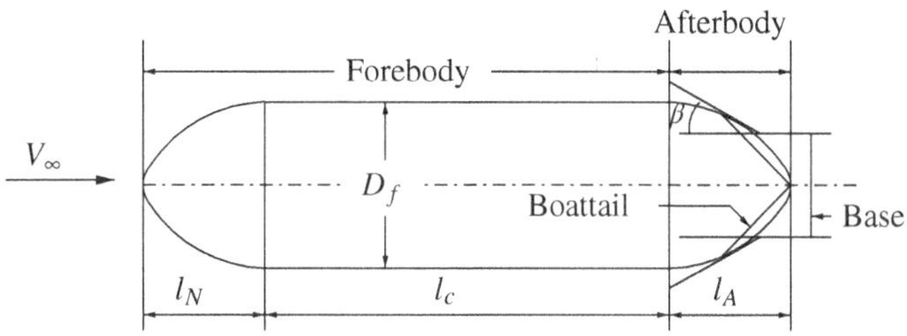

Figure 4.3 A slender body in a freestream.

or

$$\lambda_{\text{eff}} = \frac{l_N + l_A}{\sqrt{(4/\pi)A_c}} + 2 \tag{4.3b}$$

The lesser among the ones in Eqs. (4.3*a*) and (4.3*b*) will be taken as the slenderness ratio.

4.2.2 Shape Factor

The shape factor for bodies with pointed tail sections is given by

$$\phi = \frac{2.2}{\lambda_{\text{eff}}^{1.5}} + \frac{3.8}{\lambda_{\text{eff}}^{3}} \tag{4.4}$$

with the condition that

$$\frac{l_A}{\sqrt{(4/\pi)A_c}} \geq 2$$

Then, the profile drag of the slender body is obtained from Eq. (4.1) using Eqs. (4.2)–(4.4). The shape factor usually takes values between 0 and 0.5.

For bodies with blunt or short after bodies, the basic drag can be calculated from Eq. (4.1) using the lesser of the λ_{eff} defined by the following relations:

$$\lambda_{\text{eff}} = \frac{l_N + l_c}{\sqrt{(4/\pi)A_c}} + 2 \tag{4.5a}$$

$$\lambda_{\text{eff}} = 4 + \frac{l_N}{\sqrt{(4/\pi)A_c}} \tag{4.5b}$$

4.2.3 Blunt Afterbody Correction

The drag of the slender body is given by the total drag expression

$$C_{D_A} S = C_f(1 + \phi)S_{\text{wet}} + (C_{D_b})_{\text{ref}} A_c \tag{4.6}$$

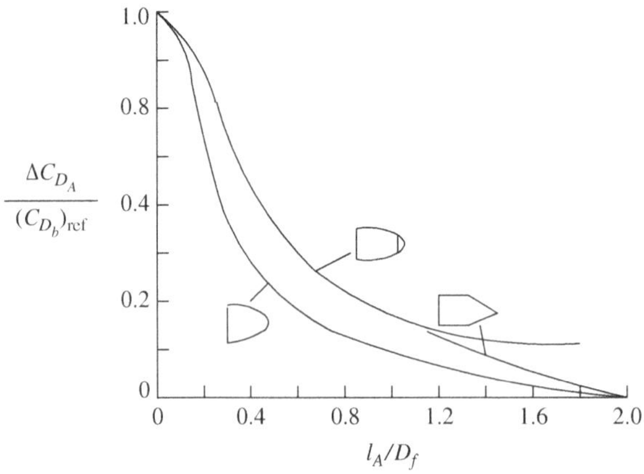

Figure 4.4 Variation of $\Delta C_{D_A}/(C_{D_b})_{ref}$ with l_A/D_f.

where

$$(C_{D_b})_{ref} = 0.015\sqrt{\frac{\sqrt{(4/\pi)A_c}}{C_f(l_N + l_C)}}$$

Variation of $\Delta C_{D_A}/(C_{D_b})_{ref}$ with l_A/D_f for some slender body shapes is shown in Fig. 4.4.

4.3 Axisymmetric Slender Bodies

For axisymmetric slender bodies, there exists a potential flow solution. This solution can yield the values of the lift, drag, and side forces acting on the body for the given potential flow assumptions. This means that the skin friction and separation are ignored.

4.3.1 Potential Flow Solution

The three forces acting on the body are given by

$$F_x = \int_0^1 \int_0^{2\pi} R'(x)R(x)\, p(x,\theta)\, d\theta\, dx$$

$$F_y = -\int_0^1 \int_0^{2\pi} R(x)\, p(x,\theta)\, \cos\theta\, d\theta\, dx$$

$$F_z = \int_0^1 \int_0^{2\pi} R(x)\, p(x,\theta)\, \sin\theta\, d\theta\, dx$$

where F_x is the drag, F_y is the side force, F_z is the lift, R is the local radius of the body at x, and R' is its first derivative.

4.3.2 Pressure Distribution

The pressure distribution on the surface of the body is given by

$$p(x, \theta) = \frac{1}{2}\rho_\infty V_\infty^2 C_p(x, \theta) + p_\infty$$

The pressure coefficient is given by

$$C_p(x, \theta) = -\frac{2q_{xA}}{V_\infty} - (R')^2 - 4\alpha_f R' \sin\theta + \alpha_f^2(1 - 4\cos\theta)$$

where

$$q_{xA} = \frac{V_\infty}{2\pi} S''(x) \ln\frac{R}{2} + \frac{V_\infty}{4\pi} S'''(x) \ln|x - x_0|\, dx_0$$

here S is the cross-sectional area of the body, and it is a function of x.

4.4 Summary

One of the typical slender bodies we see commonly is wings. Wings are only one of the types of slender body that can be tested in wind tunnel. Like wings that play a crucial role in aeronautical applications (as well as in some non-aeronautical applications), other slender bodies, such as aircraft fuselages and engine nacelles, also play equally crucial role. These bodies are not designed to produce lift although they can do. Their main aerodynamic contribution is drag. Some of the well-known slender bodies for non-aeronautical use are street luminaries, wind turbine nacelles, and anemometers.

Slender bodies are usually designed for operation at minimum drag and zero lift. Therefore, the main consideration in the design of slender bodies is the estimation of the drag of such bodies.

The best approach for estimating the drag of slender bodies is to carry out wind tunnel experiments.

Drag components consist of skin friction drag, profile drag, induced drag, and interference drag. For slender bodies, the main drag sources are the skin friction drag and profile drag. Slender bodies being streamlined shapes, the profile drag is lower than the skin friction drag. Also, drag due to interference and compressibility effects may be ignored.

The drag of fuselages, tail booms, and other slender bodies can be estimated using the flat plate analogy. The skin friction drag of the body is approximated by that of a flat plate with the same wetted area.

For a laminar flow, the skin friction coefficient of a flat plate is given by

$$C_f = \frac{1.328}{\sqrt{(\mathrm{Re})_x}}$$

For a turbulent flow, the skin friction coefficient is,

$$C_f = \frac{0.072}{[(\text{Re})_x]^{1/5}}$$

This expression is valid for Reynolds number up to 10^7, but the experimental results indicate that a better approximation is given by

$$C_f = \frac{0.074}{[(\text{Re})_x]^{1/5}}$$

Prandtl has suggested subtracting the length of the laminar layer, resulting in the expression,

$$C_f = \frac{0.074}{[(\text{Re})_x]^{1/5}} - \frac{1700}{(\text{Re})_x}$$

which can be applied for $(\text{Re})_x$ from (5×10^5) to 10^7.

For streamlined bodies, the separated zone being small, the major portion of the drag is from skin friction. Turbulent boundary layer results in more skin friction than a laminar one.

A body for which the skin friction drag is the major portion of the total drag is termed as *streamlined body*. A body for which the pressure (form) drag is the major portion is termed as *bluff body*.

The slenderness ratio of the body is defined as

$$\lambda_{\text{eff}} = \frac{l}{\sqrt{(4/\pi)A_c}}$$

or

$$\lambda_{\text{eff}} = \frac{l_N + l_A}{\sqrt{(4/\pi)A_c}} + 2$$

The shape factor for bodies with pointed tail sections is given by

$$\phi = \frac{2.2}{\lambda_{\text{eff}}^{1.5}} + \frac{3.8}{\lambda_{\text{eff}}^{3}}$$

For axisymmetric slender bodies, there exists a potential flow solution. This solution can yield the values of the lift, drag, and side force acting on the body for the given potential flow assumptions. This means that the skin friction and separation are ignored.

The pressure distribution on the surface of the body is given by

$$p(x, \theta) = \frac{1}{2}\rho_\infty V_\infty^2 C_p(x, \theta) + p_\infty$$

The pressure coefficient is given by

$$C_p(x, \theta) = -\frac{2q_{xA}}{V_\infty} - (R')^2 - 4\alpha_f R' \sin\theta + \alpha_f^2(1 - 4\cos\theta)$$

Exercise Problems

4.1 A slender body is cruising at Mach 7.5 at 3000 m altitude. Determine the skin friction coefficient experienced by the body at a distance of 3 m from its leading edge.

[Ans. 0.00177]

4.2 If a flat plate of length 1.9 m and width 1 m flies at 5.3 m/s at sea level, determine the skin friction drag acting on the plate.

[Ans. 1.47 N]

4.3 Find the drag experienced by a slender body of elliptical cross section with major axis 10 m, minor axis 1 m, and length 12 m, flying at 200 m/s at sea level, using flat plate analogy.

[Ans. 46972.135 N]

4.4 An axisymmetric cone-cylinder wind tunnel test model is 4 m long. The wind tunnel draws atmospheric air at 300 K, and runs with a test-section speed of 50 m/s. Determine the skin friction coefficient of the model.

[Ans. 0.0027]

A

Appendix

Table A.1 SI units and their conversion to US units.

	SI	US
Length	meter (m)	foot (ft) = 0.3048 m
		yard (yd) = 0.9144 m
	centimeter (cm) = 10^{-2} m	inch (in) = 25.4 mm
	kilometer (km) = 10^3 m	mile (mi) = 1.609 km
Area	square meter (m^2)	square foot (ft^2) = 0.0929 m^2
		square yard (yd^2) = 0.8361 m^2
	square centimeter (cm^2) = 10^{-4} m^2	square inch (in^2) = 6.452 cm^2
	hectare (ha) = 10^4 m^2	acre (A or ac) = 43,560 ft^2 = 0.4047 ha.
	square kilometer (km^2) =	square mile (mi^2) = 640 A = 2.590 km^2
	10^6 m^2 = 100 ha	
Volume	liter (l or L)	gallon (gal) = 3.785 l or l
		cubic foot ft^3 = 7.481 gal = 28.31 l
	cubic centimeter (cm^3) = 1 milliliter (ml)	cubic inch (in^3 = 16.39 cm^3)
	cubic meter m^3 = 10^3 l	cubic yard (yd^3) = 0.7645 m^3
Temperature	degree Celsius (°C)	degree Fahrenheit (°F) = 9/5 (°C) + 32
Mass	kilogram (kg)	pound (lb) = 0.4536 kg
	gram (g) = 10^{-3} kg	ounce (oz) = 1/16 lb = 28.35 g
	metric tonne or metric ton (t)	short ton (commonly called "ton") =
	= 10^3 kg	2000 lb = 0.9071 t
		long ton = 2240 lb = 1.016 t
	quintal (q) = 10^2 kg	hundredweight (cwt) = 112 lb = 5080.23 g

(Continued)

Hypersonic Slender Body Aerodynamics, First Edition. Ethirajan Rathakrishnan.

Companion Website: www.wiley.com/go/rathakrishnan/hypersonic

Table A.1 (Continued)

	SI	US
Force/weight	newton (N)	pound (lb) = 4.448 N
Pressure	pascal (Pa) = 1 N/m^2	
	kilopascal (kPa) = 10^3 Pa	pounds per square inch (psi) = 6.893 kPa
	torricelli (torr) = 1 mm of mercury (mm Hg) = 0.1333 kPa	inches of mercury (in Hg) = 25.4 mm Hg = 0.491 psi
Energy	joule (J)	foot-pound (ft-lb) = 1.356 J
	calorie (cal) = 4.187 J	
	kilocalorie (kcal or Cal) = 10^3 cal	British thermal unit (Btu) = 1055 J = 0.252 kcal
Power	watt (W) = joules per second (J/s)	
	kilowatt (kW) = 10^3 W	horsepower (hp) = 0.7457 kW
Moment of force or torque	N-m	pound-foot ((lb-ft) = 1.3558 N-m)

Table A.2 Properties of standard atmosphere.

h (m)	T (K)	p (kPa)	ρ (kg/m^3)	a (m/s)
0	288.150	1.01320 e+02	1.22500 e+00	3.40263 e+02
100	287.500	1.00125 e+02	1.21329 e+00	3.39879 e+02
200	286.850	9.89416 e+01	1.20167 e+00	3.39494 e+02
300	286.200	9.77695 e+01	1.19013 e+00	3.39109 e+02
400	285.550	9.66086 e+01	1.17867 e+00	3.38724 e+02
500	284.900	9.54589 e+01	1.16730 e+00	3.38338 e+02
600	284.250	9.43204 e+01	1.15602 e+00	3.37952 e+02
700	283.600	9.31928 e+01	1.14482 e+00	3.37566 e+02
800	282.950	9.20762 e+01	1.13370 e+00	3.37178 e+02
900	282.300	9.09704 e+01	1.12266 e+00	3.36791 e+02
1000	281.650	8.98755e+01	1.11171e+00	3.36403e+02
1100	281.000	8.87912e+01	1.10084e+00	3.36015e+02
1200	280.350	8.77175e+01	1.09005e+00	3.35626e+02
1300	279.700	8.66544e+01	1.07934e+00	3.35236e+02
1400	279.050	8.56017e+01	1.06871e+00	3.34847e+02
1500	278.400	8.45594e+01	1.05816e+00	3.34456e+02
1600	277.750	8.35274e+01	1.04769e+00	3.34066e+02
1700	277.100	8.25056e+01	1.03730e+00	3.33675e+02
1800	276.450	8.14940e+01	1.02700e+00	3.33283e+02
1900	275.800	8.04924e+01	1.01676e+00	3.32891e+02
2000	275.150	7.95008e+01	1.00661e+00	3.32499e+02
2100	274.500	7.85191e+01	9.96535e−01	3.32106e+02
2200	273.850	7.75473e+01	9.86537e−01	3.31712e+02

Table A.2 (Continued)

h (m)	*T* (K)	*p* (kPa)	*ρ* (kg/m^3)	*a* (m/s)
2300	273.200	7.65852e+01	9.76616e−01	3.31318e+02
2400	272.550	7.56329e+01	9.66772e−01	3.30924e+02
2500	271.900	7.46901e+01	9.57003e−01	3.30529e+02
2600	271.250	7.37569e+01	9.47310e−01	3.30134e+02
2700	270.600	7.28331e+01	9.37693e−01	3.29738e+02
2800	269.950	7.19187e+01	9.28150e−01	3.29342e+02
2900	269.300	7.10136e+01	9.18681e−01	3.28945e+02
3000	268.650	7.01178e+01	9.09287e−01	3.28548e+02
3100	268.000	6.92312e+01	8.99967e−01	3.28150e+02
3200	267.350	6.83536e+01	8.90719e−01	3.27752e+02
3300	266.700	6.74851e+01	8.81545e−01	3.27353e+02
3400	266.050	6.66255e+01	8.72443e−01	3.26954e+02
3500	265.400	6.57749e+01	8.63413e−01	3.26554e+02
3600	264.750	6.49330e+01	8.54455e−01	3.26154e+02
3700	264.100	6.40999e+01	8.45568e−01	3.25754e+02
3800	263.450	6.32754e+01	8.36752e−01	3.25352e+02
3900	262.800	6.24596e+01	8.28006e−01	3.24951e+02
4000	262.150	6.16523e+01	8.19330e−01	3.24549e+02
4100	261.500	6.08535e+01	8.10724e−01	3.24146e+02
4200	260.850	6.00631e+01	8.02188e−01	3.23743e+02
4300	260.200	5.92810e+01	7.93720e−01	3.23339e+02
4400	259.550	5.85071e+01	7.85321e−01	3.22935e+02
4500	258.900	5.77415e+01	7.76990e−01	3.22531e+02
4600	258.250	5.69840e+01	7.68727e−01	3.22126e+02
4700	257.600	5.62346e+01	7.60531e−01	3.21720e+02
4800	256.950	5.54931e+01	7.52402e−01	3.21314e+02
4900	256.300	5.47596e+01	7.44340e−01	3.20907e+02
5000	255.650	5.40340e+01	7.36344e−01	3.20500e+02
5100	255.000	5.33162e+01	7.28414e−01	3.20092e+02
5200	254.350	5.26061e+01	7.20550e−01	3.19684e+02
5300	253.700	5.19037e+01	7.12750e−01	3.19275e+02
5400	253.050	5.12089e+01	7.05016e−01	3.18866e+02
5500	252.400	5.05217e+01	6.97345e−01	3.18456e+02
5600	251.750	4.98419e+01	6.89739e−01	3.18046e+02
5700	251.100	4.91696e+01	6.82196e−01	3.17635e+02
5800	250.450	4.85046e+01	6.74717e−01	3.17224e+02
5900	249.800	4.78469e+01	6.67300e−01	3.16812e+02
6000	249.150	4.71965e+01	6.59946e−01	3.16399e+02
6100	248.500	4.65532e+01	6.52654e−01	3.15986e+02
6200	247.850	4.59171e+01	6.45424e−01	3.15573e+02
6300	247.200	4.52880e+01	6.38255e−01	3.15159e+02
6400	246.550	4.46659e+01	6.31148e−01	3.14744e+02
6500	245.900	4.40508e+01	6.24101e−01	3.14329e+02

(Continued)

Table A.2 (Continued)

h (m)	T (K)	p (kPa)	ρ (kg/m^3)	a (m/s)
6600	245.250	4.34425e+01	6.17114e−01	3.13913e+02
6700	244.600	4.28411e+01	6.10187e−01	3.13497e+02
6800	243.950	4.22464e+01	6.03320e−01	3.13080e+02
6900	243.300	4.16584e+01	5.96513e−01	3.12663e+02
7000	242.650	4.10770e+01	5.89764e−01	3.12245e+02
7100	242.000	4.05023e+01	5.83074e−01	3.11826e+02
7200	241.350	3.99340e+01	5.76442e−01	3.11407e+02
7300	240.700	3.93723e+01	5.69867e−01	3.10988e+02
7400	240.050	3.88169e+01	5.63351e−01	3.10567e+02
7500	239.400	3.82679e+01	5.56891e−01	3.10147e+02
7600	238.750	3.77252e+01	5.50488e−01	3.09725e+02
7700	238.100	3.71888e+01	5.44142e−01	3.09303e+02
7800	237.450	3.66586e+01	5.37852e−01	3.08881e+02
7900	236.800	3.61345e+01	5.31618e−01	3.08458e+02
8000	236.150	3.56164e+01	5.25439e−01	3.08034e+02
8100	235.500	3.51044e+01	5.19315e−01	3.07610e+02
8200	234.850	3.45984e+01	5.13246e−01	3.07185e+02
8300	234.200	3.40983e+01	5.07231e−01	3.06760e+02
8400	233.550	3.36041e+01	5.01270e−01	3.06334e+02
8500	232.900	3.31157e+01	4.95363e−01	3.05907e+02
8600	232.250	3.26330e+01	4.89510e−01	3.05480e+02
8700	231.600	3.21561e+01	4.83709e−01	3.05052e+02
8800	230.950	3.16848e+01	4.77961e−01	3.04624e+02
8900	230.300	3.12191e+01	4.72266e−01	3.04195e+02
9000	229.650	3.07590e+01	4.66623e−01	3.03765e+02
9100	229.000	3.03044e+01	4.61031e−01	3.03335e+02
9200	228.350	2.98553e+01	4.55491e−01	3.02904e+02
9300	227.700	2.94115e+01	4.50002e−01	3.02473e+02
9400	227.050	2.89731e+01	4.44563e−01	3.02041e+02
9500	226.400	2.85400e+01	4.39175e−01	3.01608e+02
9600	225.750	2.81122e+01	4.33837e−01	3.01175e+02
9700	225.100	2.76896e+01	4.28549e−01	3.00741e+02
9800	224.450	2.72721e+01	4.23311e−01	3.00307e+02
9900	223.800	2.68598e+01	4.18121e−01	2.99871e+02
10000	223.150	2.64525e+01	4.12980e−01	2.99436e+02
10100	222.500	2.60502e+01	4.07888e−01	2.98999e+02
10200	221.850	2.56529e+01	4.02844e−01	2.98562e+02
10300	221.200	2.52605e+01	3.97848e−01	2.98124e+02
10400	220.550	2.48730e+01	3.92899e−01	2.97686e+02
10500	219.900	2.44903e+01	3.87997e−01	2.97247e+02
10600	219.250	2.41124e+01	3.83143e−01	2.96807e+02
10700	218.600	2.37392e+01	3.78335e−01	2.96367e+02
10800	217.950	2.33708e+01	3.73573e−01	2.95926e+02
10900	217.300	2.30069e+01	3.68857e−01	2.95485e+02

Table A.2 (Continued)

h (m)	T (K)	p (kPa)	ρ (kg/m^3)	a (m/s)
11100	216.650	2.22935e+01	3.58492e−01	2.95042e+02
11200	216.650	2.19449e+01	3.52886e−01	2.95042e+02
11300	216.650	2.16018e+01	3.47368e−01	2.95042e+02
11400	216.650	2.12640e+01	3.41936e−01	2.95042e+02
11500	216.650	2.09314e+01	3.36589e−01	2.95042e+02
11600	216.650	2.06041e+01	3.31326e−01	2.95042e+02
11700	216.650	2.02819e+01	3.26145e−01	2.95042e+02
11800	216.650	1.99648e+01	3.21044e−01	2.95042e+02
11900	216.650	1.96526e+01	3.16024e−01	2.95042e+02
12000	216.650	1.93453e+01	3.11082e−01	2.95042e+02
12100	216.650	1.90427e+01	3.06218e−01	2.95042e+02
12200	216.650	1.87450e+01	3.01429e−01	2.95042e+02
12300	216.650	1.84518e+01	2.96716e−01	2.95042e+02
12400	216.650	1.81633e+01	2.92076e−01	2.95042e+02
12500	216.650	1.78793e+01	2.87508e−01	2.95042e+02
12600	216.650	1.75997e+01	2.83013e−01	2.95042e+02
12700	216.650	1.73245e+01	2.78587e−01	2.95042e+02
12800	216.650	1.70536e+01	2.74230e−01	2.95042e+02
12900	216.650	1.67869e+01	2.69942e−01	2.95042e+02
13000	216.650	1.65244e+01	2.65721e−01	2.95042e+02
13100	216.650	1.62660e+01	2.61566e−01	2.95042e+02
13200	216.650	1.60116e+01	2.57476e−01	2.95042e+02
13300	216.650	1.57612e+01	2.53449e−01	2.95042e+02
13400	216.650	1.55148e+01	2.49486e−01	2.95042e+02
13500	216.650	1.52722e+01	2.45585e−01	2.95042e+02
13600	216.650	1.50333e+01	2.41744e−01	2.95042e+02
13700	216.650	1.47983e+01	2.37964e−01	2.95042e+02
13800	216.650	1.45668e+01	2.34243e−01	2.95042e+02
13900	216.650	1.43391e+01	2.30580e−01	2.95042e+02
14000	216.650	1.41148e+01	2.26974e−01	2.95042e+02
14100	216.650	1.38941e+01	2.23425e−01	2.95042e+02
14200	216.650	1.36768e+01	2.19931e−01	2.95042e+02
14300	216.650	1.34630e+01	2.16492e−01	2.95042e+02
14400	216.650	1.32524e+01	2.13106e−01	2.95042e+02
14500	216.650	1.30452e+01	2.09774e−01	2.95042e+02
14600	216.650	1.28412e+01	2.06494e−01	2.95042e+02
14700	216.650	1.26404e+01	2.03265e−01	2.95042e+02
14800	216.650	1.24427e+01	2.00086e−01	2.95042e+02
14900	216.650	1.22482e+01	1.96957e−01	2.95042e+02
15000	216.650	1.20566e+01	1.93877e−01	2.95042e+02
15100	216.650	1.18681e+01	1.90846e−01	2.95042e+02
15200	216.650	1.16825e+01	1.87861e−01	2.95042e+02
15300	216.650	1.14998e+01	1.84924e−01	2.95042e+02
15400	216.650	1.13200e+01	1.82032e−01	2.95042e+02

(Continued)

Table A.2 (Continued)

h (m)	T (K)	p (kPa)	ρ (kg/m^3)	a (m/s)
15500	216.650	1.11430e+01	1.79185e−01	2.95042e+02
15600	216.650	1.09687e+01	1.76383e−01	2.95042e+02
15700	216.650	1.07972e+01	1.73625e−01	2.95042e+02
15800	216.650	1.06284e+01	1.70910e−01	2.95042e+02
15900	216.650	1.04622e+01	1.68237e−01	2.95042e+02
16000	216.650	1.02986e+01	1.65607e−01	2.95042e+02
16100	216.650	1.01375e+01	1.63017e−01	2.95042e+02
16200	216.650	9.97900e+00	1.60468e−01	2.95042e+02
16300	216.650	9.82295e+00	1.57958e−01	2.95042e+02
16400	216.650	9.66934e+00	1.55488e−01	2.95042e+02
16500	216.650	9.51814e+00	1.53057e−01	2.95042e+02
16600	216.650	9.36930e+00	1.50663e−01	2.95042e+02
16700	216.650	9.22279e+00	1.48307e−01	2.95042e+02
16800	216.650	9.07857e+00	1.45988e−01	2.95042e+02
16900	216.650	8.93660e+00	1.43705e−01	2.95042e+02
17000	216.650	8.79685e+00	1.41458e−01	2.95042e+02
17100	216.650	8.65929e+00	1.39246e−01	2.95042e+02
17200	216.650	8.52388e+00	1.37069e−01	2.95042e+02
17300	216.650	8.39059e+00	1.34925e−01	2.95042e+02
17400	216.650	8.25938e+00	1.32815e−01	2.95042e+02
17500	216.650	8.13023e+00	1.30738e−01	2.95042e+02
17600	216.650	8.00309e+00	1.28694e−01	2.95042e+02
17700	216.650	7.87794e+00	1.26682e−01	2.95042e+02
17800	216.650	7.75475e+00	1.24701e−01	2.95042e+02
17900	216.650	7.63348e+00	1.22751e−01	2.95042e+02
18000	216.650	7.51412e+00	1.20831e−01	2.95042e+02
18100	216.650	7.39661e+00	1.18942e−01	2.95042e+02
18200	216.650	7.28095e+00	1.17082e−01	2.95042e+02
18300	216.650	7.16709e+00	1.15251e−01	2.95042e+02
18400	216.650	7.05502e+00	1.13449e−01	2.95042e+02
18500	216.650	6.94470e+00	1.11675e−01	2.95042e+02
18600	216.650	6.83610e+00	1.09928e−01	2.95042e+02
18700	216.650	6.72920e+00	1.08209e−01	2.95042e+02
18800	216.650	6.62397e+00	1.06517e−01	2.95042e+02
18900	216.650	6.52039e+00	1.04851e−01	2.95042e+02
19000	216.650	6.41843e+00	1.03212e−01	2.95042e+02
19100	216.650	6.31806e+00	1.01598e−01	2.95042e+02
19200	216.650	6.21926e+00	1.00009e−01	2.95042e+02
19300	216.650	6.12200e+00	9.84452e−02	2.95042e+02
19400	216.650	6.02627e+00	9.69058e−02	2.95042e+02
19500	216.650	5.93204e+00	9.53904e−02	2.95042e+02
19600	216.650	5.83927e+00	9.38987e−02	2.95042e+02
19700	216.650	5.74796e+00	9.24304e−02	2.95042e+02
19800	216.650	5.65808e+00	9.09850e−02	2.95042e+02
19900	216.650	5.56960e+00	8.95622e−02	2.95042e+02
20000	216.650	5.48251e+00	8.81617e−02	2.95042e+02

References

1 E. Rathakrishnan, *Applied Gas Dynamics*, 2nd ed. John Wiley & Sons, Inc., Hoboken, NJ, 2019.
2 S. Rudman and S. G. Rubin, "Hypersonic viscous flow over slender bodies with sharp leading edges", *AIAA J.*, Vol. 6, No. 10, pp. 1883–1890, 1968, doi: 10.2514/3.4896.
3 Q.-X. Wang, "Analysis of slender bodies of revolution with curved-ground effect and waving-water effect", *Fluid Dyn. Res.*, Vol. 9, No. 5-6, pp. 235–254, 1992, doi: 10.1016/0169-5983(92)90041-T.
4 C. F. Dewey, Jr., "Bluntness and Viscous-Interaction Effects on Slender Bodies at Hypersonic Speeds", Memorandum RM-3832-PR, The RAND Corporation Santa Monica - California, Unites States, Sept. 1964.
5 H. W. Liepmann and A. Roshko, *Elements of Gasdynamics*, John Wiley & Sons, Inc., New York, 1957.
6 K. W. Mangler and J. H. B. Smith, "Calculations of the flow past slender delta wings leading edge separation", RAE Report Aero 2593, 1957.
7 K. Oswatitsch, "Nichlineare Tragflächentheorie, insbesondere fü Tragfü mit kleinem Seitenverhältnis", *Ing. Arch.*, Vol. 31, pp. 431–452, 1961.
8 R. Hürlimann, "Zur Berechnung der Kräfte auf schlanke, zugespitzte Tragflü", *Z. Flugwiss.*, Vol. 16, pp. 69–81, 1961.
9 G. Koppenwallner, "Experimentally Untersuchung der Druckverteilung und des Widerstands von querangestroemten Kreiszylindern bei hypersonischen Machzahlen in Bereich von Kontinuums – bis freier Molekularstro- mung", *Z. Flugwiss.*, Vol. 17, No. 10, pp. 321–332, Oct. 1969.
10 Anon, "Standard Atmosphere", ISO 2533:1975, International Organization for Standardization, 1975.
11 Anon, "U.S. Standard Atmosphere 1976", NASA-TM-X-74335; NOAAS/T-76-1562, 1976.
12 E. Rathakrishnan, *High Enthalpy Gas Dynamics*, John Wiley & Sons, Inc., Hoboken, NJ, 2015.

Hypersonic Slender Body Aerodynamics, First Edition. Ethirajan Rathakrishnan.

Companion Website: www.wiley.com/go/rathakrishnan/hypersonic

13 J. D. Cole and J. Aroesty, "Hypersonic similarity solutions for airfoils supporting exponential shock waves", RAND Corporation, Memorandum RM - PR, 1968.

14 A. Miele (Editor), *Theory of Optimum Aerodynamic Shapes*, Academic Press, New York/London, 1965.

15 J. Seddon and A. Spence, "The use of known flow fields as an approach to the design of high speed aircraft. Hypersonic Boundary Layers and Flow Fields", AGARD Conference Proceedings No. 30, May 1968.

16 J. G. Jones, K. C. Moore, J. Pike and P. L. Roe, "A method for designing lifting configurations for high supersonic speeds, using axisymmetric flow fields", *Ing. Arch.*, Vol. 37, pp. 56–72, 1968.

17 A. F. Messiter, "Lift of slender delta wings according to Newtonian theory", *AIAA J.*, Vol. 1, pp. 794–802, 1963.

18 K. Hida, "Thickness effect on the force of slender delta wings in hypersonic flow", *AIAA J.*, Vol. 3, pp. 427–433, 1965.

19 L. C. Squire, "Calculated pressure distribution and shock shapes on conical wings at high supersonic speeds", *Aeronaut. Q.*, Vol. 18, pp. 185–206, 1967.

20 L. C. Squire, "Calculated pressure distribution and shock shapes on conical wings with attached shock waves", *Aeronaut. Q.*, Vol. 19, pp. 31–50, 1968.

21 W. D. Hayes and R. F. Probstein, *Hypersonic Flow Theory*, Vol. I Inviscid Flows, Academic Press, New York, 1966.

22 E. Rathakrishnan, *Fluid Mechanics: An Introduction*, 4th ed. PHI Learning, Delhi, 2022.

23 H. Schlichting and K. Gersten, *Boundary-Layer Theory*, 9th ed. Springer, 2017.

Further Readings

H. Ashley and M. Landahl, *Aerodynamics of Wings and Bodies*, Dover Publications, Inc., New York, 1965.

J. J. Bertin, *Hypersonic Aerothermodynamics*. AIAA Education Series, AIAA, 1994.

J. C. Cooke, "Properties of a Two-Parameter Family of Thin Conically Cambered Delta Wings by Slender-Body Theory, R. & M. No. 3249", Her Majesty's Stationery Office, London, 1962.

R. N. Cox and L. F. Crabtree, *Elements of Hypersonic Aerodynamics*, Academic Press, New York, 1965.

S. F. Hoerner, *Fluid-Dynamic Drag*, Published by the Author, 1965.

L. H. Jorgensen, "A method for estimating static aerodynamic characteristics for slender bodies of circular and noncircular cross section alone and with lifting surfaces at angles of attack from 0° to 90°", NASA-TN-D-7228, 1973.

L. H. Jorgensen and E. R. Nelson, "Experimental aerodynamics characteristics for a cylindrical body of revolution with side strakes and various noses at angles of attack from 0° to 58°and Mach numbers from 0.6 to 2.0", NASA-TM-X-3130, 1975.

E. S. Levinsky and M. H. Y. Wei, "Non-linear lift and pressure distribution of slender conical bodies with strakes at low speeds", NASA CR-1202, 1968.

R. J. Muraca, "An empirical method for determining static distributed aerodynamic loads on axisymmetric multistage launch vehicles", NASA TN D-3283, 1966.

M. Rasmussen, *Hypersonic Flow*, John Wiley & Sons, New York, 1994

E. Rathakrishnan, *Gas Dynamics*, 7th ed. PHI Learning, Delhi, 2021.

J. L. Sims, "Tables for supersonic flow around right circular cones at small angles of attack", NASA SP-3007, 1964.

Hypersonic Slender Body Aerodynamics, First Edition. Ethirajan Rathakrishnan.

Companion Website: www.wiley.com/go/rathakrishnan/hypersonic

Index

Hypersonic Slender Body Aerodynamics, First Edition. Ethirajan Rathakrishnan.

Companion Website: www.wiley.com/go/rathakrishnan/hypersonic